Pseudomonas aeruginosa - New Perspectives and Applications

Edited by Osama Darwesh and Ibrahim Matter

Published in London, United Kingdom

Pseudomonas aeruginosa - New Perspectives and Applications
http://dx.doi.org/10.5772/intechopen.100897
Edited by Osama Darwesh and Ibrahim Matter

Contributors
Amalia Muñoz-Gómez, Anika Farzin, Diana Carolina Castaño, Ehsan F. F. Hussein, Ferdows Ara Mollika, Juan David Patiño-Salazar, Mahir Ahmed, Mauricio Corredor, Md. Mizanur Rahman, Mohammad Sharif Sarker, Nirmal Chandra Barman, Sankar Ramachandran, Sonal Malhotra, Sunil Kumar, Zahur Ahmed

First published in London, United Kingdom, 2024 by IntechOpen
IntechOpen is the global imprint of INTECHOPEN LIMITED, registered in England and Wales, registration number: 11086078, 167-169 Great Portland Street, London, W1W 5PF, United Kingdom

British Library Cataloguing-in-Publication Data
A catalogue record for this book is available from the British Library

Additional hard and PDF copies can be obtained from orders@intechopen.com

Pseudomonas aeruginosa - New Perspectives and Applications
Edited by Osama Darwesh and Ibrahim Matter
p. cm.
Print ISBN 978-1-80356-824-9
Online ISBN 978-1-80356-825-6
eBook (PDF) ISBN 978-1-80356-826-3

For EU product safety concerns:
IN TECH d.o.o., Prolaz Marije Krucifikse Kozulić 3, 51000 Rijeka, Croatia,
info@intechopen.com or visit our website at intechopen.com.

Meet the editors

Prof. Osama Darwesh is a Professor of Environmental Microbiology and Nanotechnology in the Agricultural Microbiology Department, National Research Centre, Cairo, Egypt. He has been awarded 6 prizes including the National Research Center's Best Ph.D. Thesis award in 2014, the Environmental Research Award from the Academy of Scientific Research and Technology in 2017, and the Encouraging State Prize in Advanced Technological Sciences in 2018. He holds 6 patents and has published more than 110 international papers and 15 books.

Prof. Ibrahim Matter is a Professor of Environmental Microbiology and Biotechnology at the National Research Centre, Cairo, Egypt. He obtained a Ph.D. in Industrial Microbiology from Cairo University, Egypt, in 2014. He was a visiting scientist at several universities in South Korea and Australia and has authored several book chapters on microbial biotechnology and nanotechnology.

Contents

Preface

Pseudomonas aeruginosa is an opportunistic bacterium associated with a wide range of diseases that is commonly found in soil and water. It is also the leading cause of hospital-acquired infections and is associated with high morbidity and mortality among elderly and/or immunocompromised patients. Therefore, it represents a great challenge for health workers. As such, there are many benefits to be gained from studying these bacteria from biotechnological, industrial, environmental, and medical aspects. This book, Pseudomonas aeruginosa – *New Perspectives and Applications*, is an edited collection of reviewed and relevant research chapters on *P. aeruginosa*, which has significant implications for industry, medicine, and the environment. The book includes scientific contributions from various authors, edited by a group of experts in immunology and microbiology. Each contribution is a completely separate chapter in itself but is directly linked to the themes and aims of the book.

The book includes the following chapters:

Chapter 1: "The Role of *Pseudomonas aeruginosa* DsbA-1 in Bacterial Pathogenesis: Current Research and Future Prospects"

Chapter 2: "Searching for the Resistance Interactome of *Pseudomonas aeruginosa*"

Chapter 3: "*Pseudomonas aeruginosa*: The Alarming Pathogen of Hospital Acquired Infection"

Chapter 4: "Optimizing Shelf-life of *Pseudomonas fluorescens* after Freeze Drying"

Chapter 5: "The Pangenome of *Pseudomonas aeruginosa*"

Chapter 6: "*Pseudomonas aeruginosa* Represents a Main Cause of Hospital-Acquired Infections (HAI) and Multidrug Resistance (MDR)"

This book provides comprehensive information about this multifaceted microbe to scientists and specialists in this field for the benefit of humanity.

Osama Darwesh and Ibrahim Matter
Professor,
Agricultural Microbiology Department,
National Research Centre,
Cairo, Egypt

Chapter 1

The Role of *Pseudomonas aeruginosa* DsbA-1 in Bacterial Pathogenesis: Current Research and Future Prospects

Sunil Kumar and Sonal Malhotra

Abstract

Disulfide bond isomerase proteins (Dsbs) have been extensively characterized in gram-negative bacteria. Recently research efforts is being placed on their biology in gram-positive species. Modern "omics" technologies, allowed assessment of the contribution of the Dsbs to bacterial pathogenesis. The author cloned and characterized the *dsbA* 1 protein from *Pseudomonas aeruginosa* in the late 1990s. The global proteome analysis demonstrated that the *dsbA* gene is under the direct regulatory control of the extracytoplasmic function (ECF) sigma factor AlgT(U) or sigma-22. This is unique to *P. aeruginosa*. Disruption of *dsbA* gene results in pleiotropic phenotype: defect in assembly of cysteine disulfide bond containing proteins-as shown in many others. Recently, omics-based approaches identified expression changes in *dsbA* gene under different physiological states of bacterial pathogens-primarily in free-living, biofilm state, or under infectious disease conditions. Involvement of *dsbA* function in biofilm formation was shown using *dsbA* gene disruption mutants. This chapter documents past and current findings and concludes with future trends in research on Dsbs including peptidomimetics.

Keywords: bacterial pathogenesis, disulfide isomerase, OMICS technology, proteomics, extracytoplasmic function (ECF) sigma factors, thiol status, Sigma factors, microarrays, metatranscriptomics, metabolomics, animal models of disease

1. Introduction

This chapter examines the biology of disulfide bond isomerase (Dsb) proteins and how these proteins are involved in assembly of virulence determinants in the bacterial cell [1]. We first examine the function of these proteins within the cell. DsbAs are proteins that facilitate the formation of the disulfide bond. This bond exists between two cysteine amino acid residues. The "R" group in this amino acid contains thiol or "SH" group at its end. The thiol group needs to be oxidized and DsbA or DsbA-like proteins aid in this. Slow oxidation of the thiol group is possible (**Figure 1**). The disulfide bond can be intrachain-within one polypeptide or alternatively between two

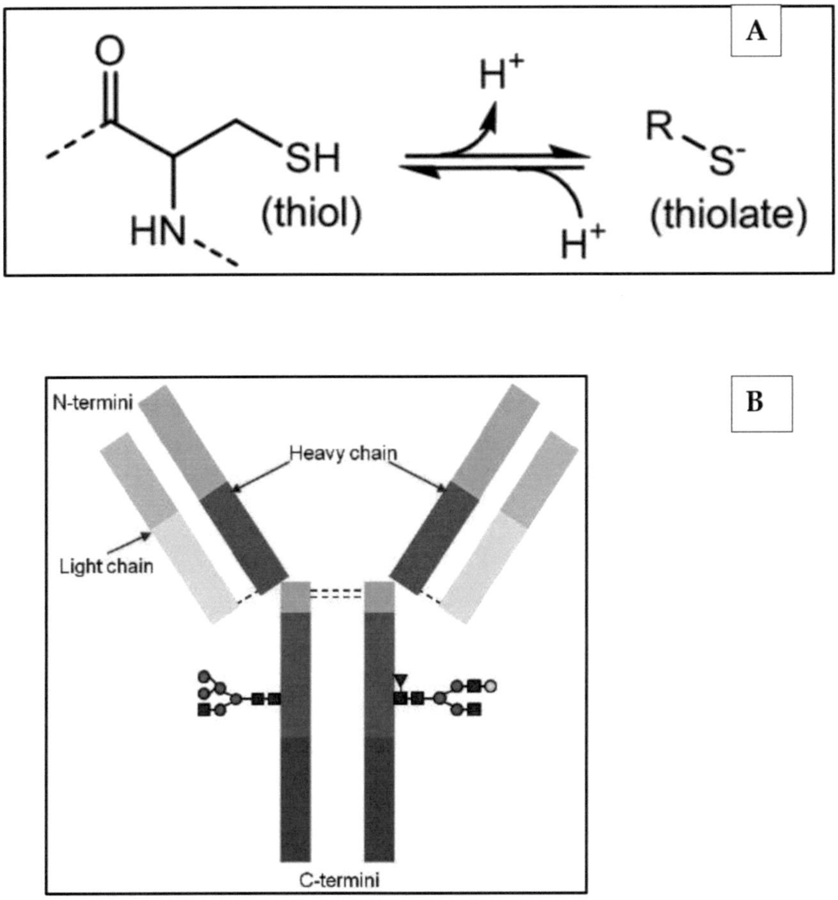

Figure 1.
SH group and disulfide bond in protein. ***Panel A****: The generation of the reactive intermediate the "thiolate" ion. This reactive center can further react to form a disulfide bond. The disulfide bond stabilizes protein structure. Most proteins are held by chaperones and maintained till their ultimate death and destruction [2]. In Panel* ***B*** *you can see both intermolecular disulfide bonds (between heavy and light chain) and intramolecular (In the heavy chain).*

polypeptide chains (intermolecular). Several kinds of proteins have disulfide bonds. The DsbA protein is localized in the periplasmic space in gram-negative bacteria. It is involved in oxidation reactions with the thiol groups in proteins. It is regenerated by a cognate DsbB protein localized in the inner membrane of the gram-negative bacterial cell [3]. In gram-positive bacteria, the role of DsbA and Dsbs is less well understood. This is likely due to the lack of known and tractable model organisms and/or the lack of genetic tools for analysis [4]. Peculiar characteristics of gram-positives will be discussed later in the text [5]. Scientists study DsbA function by disrupting the coding sequence of the *dsbA* gene. Typically this is done by insertion of a nonpolar cassette insertion that encodes a Gentamycin drug resistance marker [6]. The lack of DsbA function causes a slew of phenotypic changes in the mutant [6]. These changes are well characterized and typically similar in gram-negative bacteria [7]. Interestingly you can complement a *dsbA* disruption in *P. aeruginosa* by complementing with a cloned *E. coli dsbA* gene. Inspite of the *dsbA* mutation, the strain will survive and typically exhibits growth characteristics similar to the wild-type parent strain [6].

This property makes it easy to study the *dsbA* mutation. Similar growth rates between the mutant and wild-type strains indicate that the effects of *dsbA* mutation do not suppress growth of bacteria. Growth differences could further complicate analysis. Due to similar growth rates its likely phenotypic differences are due to *dsbA* disruption. This allows direct comparison of wild type to the mutant data set [6]. These observations are supported by complementation studies: wherein a cloned *dsbA* gene restores the defective phenotypes in the mutant [6]. This is almost true for at least all gram-negative bacteria I have come across and worked with including *Sinorhizobium meliloti, Haemophilus influenzae,* and the gram-positive bacteria *Staphylococcus aureus* [unpublished data].

2. *Pseudomonas aeruginosa*: an opportunistic pathogen

P. aeruginosa has a large genome size (6.3 Mb). It is a gram-negative rod-shaped bacterium that is often classified as an opportunistic pathogen. It is also a member of the ESKAPE pathogen group [8]. This opportunistic pathogen typically does not cause lethal disease in normal healthy human individuals. However, in people with comorbidities and other lifestyle disorders and accidents (burns surgical wounds and in-dwelling devices) *P. aeruginosa* can cause severe protracted illness and is a common cause of nosocomial or hospital-acquired infections [8]. It is a designated superbug. Thus, understanding its pathogenesis is key. Its genome (sequenced in the year 2000, Pathogenesis Corporation) is relatively more complex than *Escherichia coli* with some fascinating distinctions [9]. During the years 1996 through 2000, the author was involved in delineating the stress response regulon of AlgU/T or Sigma-22. Sigma-22 is a member of a conserved subfamily of Sigma 70 sigma factors. This subfamily of sigma factors is united through: the conservation of its protein structure; types of genes that is group regulates; the unique mechanism of activation in response to extracytoplasmic stress and potential consensus elements in the promoters of its induced genes. It's appropriately named the Extracytoplasmic function (ECF) subfamily of sigma factors—the fourth member of the Sigma 70 group of sigma factors [10]. For a long time, scientists have considered colonization of a host species to involve a stress response in the invading bacterial pathogen. The pathogen adapts to the stress and then colonizes a niche within the host. Different OMICS based technologies have been developed. Scientists are using these technologies to study the stress response pathogenic bacteria generate in response to altered host species environment they encounter. There are 19 ECF sigma factors in *P. aeruginosa*. The extent of the regulons controlled by ECF sigma factors *in P. aeruginosa* is under study. The author used a reverse genetic approach proteomics-based approach to identify targets for the well-characterized sigma factor AlgT/U or Sigma-22. AlgT was first identified as a major virulence determinant due to its involvement in activation of the alginate biosynthesis operon. Alginate is a mucous exopolysaccharide that is a major virulence determinant in CF disease pathogenesis involving *P. aeruginosa*. Its regulation was characterized by Prof. Dennis E. Ohman a Ph.D. Mentor of the author, to whom the author is highly obliged [6].

3. Alginate: the Mucous exopolysaccharide

Alginate is a mucous polysaccharide that is made of individual sugar units that are assembled in the cell and extruded out as a polymer. It is an important virulence determinant of *P. aeruginosa* [11, 12]. Especially in the context of persistent Cystic fibrosis infections, where opportunistic *P. aeruginosa* infections can and do turn lethal. It is now believed that bacterial and viral co-infections are the main reason for the greater extent of morbidity of the disease in CF patients. Novel methods are being developed to fight these infections. These will be discussed in the summary section of the chapter.

Alginate is a polymer composed of *P. aeruginosa* alginate is composed of D-mannuronic acid residues interspersed with L-guluronic acid residues. The hydroxyl groups of the D-mannuronic acid residues may be O-acetylated at the C2′ and/or C3′ positions. The polymer is unbranched and has a long chain. The alginate biosynthesis operon is about 18 Kb long and is regulated by the ECF sigma factor AlgT. The acetylation function is carried out by the proteins AlgJ and AlgX, where AlgX might be involved in sequestering the nascent alginate polymer in the periplasm. The coating of the exopolymer alginate may provide a survival advantage to cells. It reduces the intake of antibiotics and phagocytosis. Alginate also prevents desiccation and provides protection from oxidizing agents [13]. Consequently, alginate-producing isolates of *P. aeruginosa* may have a survival advantage in particular in CF patients.

The lung function decreases upon the appearance of the mucoid (Alginate + phenotype) in *P. aeruginosa* in CF lung. This also correlates with a decrease of some other bacterial species like *S. aureus* and *H. influenzae*. The mucoid phenotype is a manifestation of the production of alginate. So, we need to focus on what alginate does. Alginate can scavenge hypochlorite produced by phagocytic cells. *P. aeruginosa* alginate also reduces the chemotaxis of polymorphonuclear leukocytes (PMN) into CF lungs. Alginate innately can inhibit the activation of the complement system. *P. aeruginosa* cells and synthesizing alginate are inherently resistant to phagocytosis by both PMN and macrophages in comparison to nonmucoid strains. Additionally, binding of the bacteria to host factors such as respiratory mucins or growth in a biofilm can aid in the colonization of CF lung environment and stabilization of infection (**Figure 2**) [16].

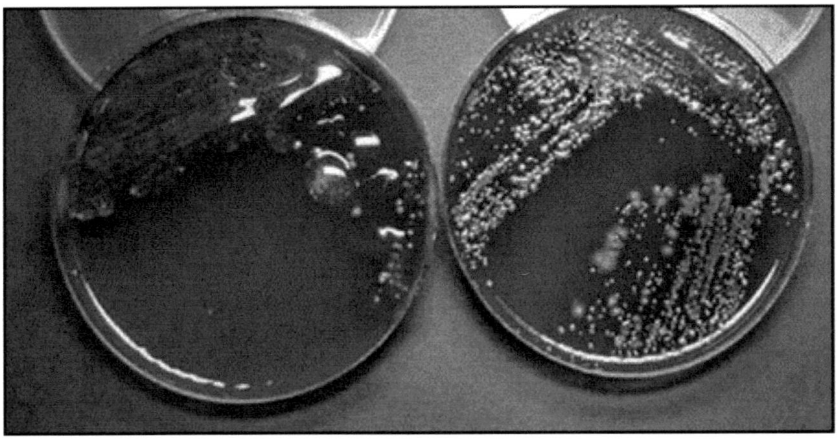

A B

Figure 2.
Colony morphology of Pseudomonas aeruginosa *on plates- mucoid phenotype. Mucoid* ***(A)*** *and nonmucoid* ***(B)*** *Colony variants of* Pseudomonas aeruginosa *on sheep blood agar plates. The mucoid phenotype is due to excess production of the polysaccharide Alginate [14, 15].*

DOI: http://dx.doi.org/10.5772/intechopen.108072

4. Elastase the well characterized secreted protease

Numerous proteases are secreted by *P. aeruginosa*. They are concentrated outside the cell (extracellular). They operate as virulence factors. The LasB and LasA elastases are the most important. Alkaline protease (AprA), type IV protease (PIV), *P. aeruginosa* small protease (PASP), large ExoProtease A (LepA), *P. aeruginosa* aminopeptidase (PAAP), and MucD are the other proteases that *P. aeruginosa* produces [14]. The T2SS transports the LasA and LasB elastases extracellular environment [17]. The Quorum sensing (QS) system regulates the expression of their genes (lasA & lasB) [14]. Elastase or pseudolysin are other names for LasB elastase. The host protein Elastin is broken down by LasB a zinc-dependent metalloprotease [17]. It is the most

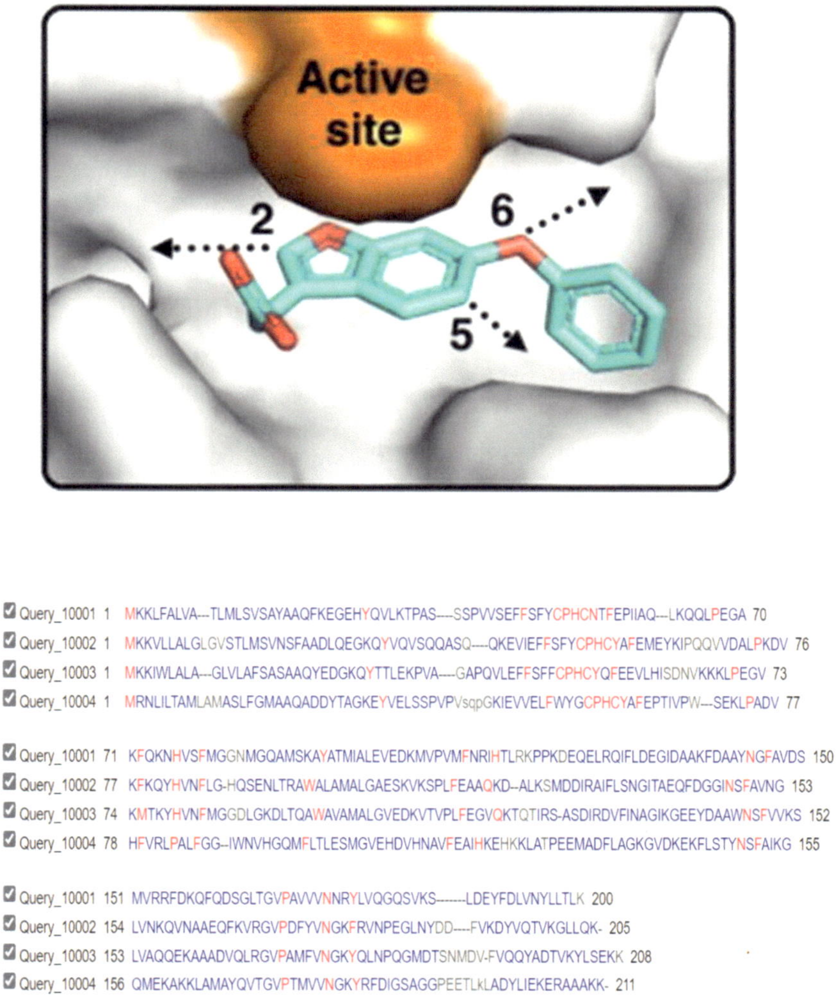

Figure 3.
Multiple sequence alignment dsbA *proteins & catalytic site. Top Panel: Graphical representation of an inhibitor designed to fit into the hydrophobic pocket/groove of DsbA. The region around the active site is highly conserved. It is believed that the sequence around the catalytic domain CPHC motif determines the reactivity of the Cysteine residues. You can see strong conservation around the CPHC motif in the MSA in Panel 2. Sequences 1 through 4 are* Vibrio Cholera, Haemophilus, E. coli, *and* Pseudomonas *DsbA proteins.*

prevalent extracellular enzyme, making it a focus of diligent research. In addition to its elastolytic activity, it also has other substrates, including immunoglobulins, surfactant proteins (SP-A and SP-D), cytokines (TNF-, IFN-, IL-6, or IL-2), and IL-2. Biofilm development is also impacted by LasB [13, 16]. We concentrate particularly on elastase because it is the most abundant enzyme. In addition, the author characterized its secretion in a *dsbA* mutant. The disulfide bonds exist in Elastase and the type II secretion system components [14]. By using a western blot and assessing the protein's activity using a common elastase activity assay, we were able to demonstrate decreased levels of the Elastase protein and reduced elastolytic activity in our DsbA characterization (**Figure 3**) [6]. The serine protease LasA or staphylolysin is produced by the lasA gene. A zinc-dependent metalloendopeptidase called alkaline protease, also known as aeruginolysin, is released through the T1SS and is made by the aprA gene [14]. Its primary protein targets are endothelium's fibronectin and laminin complement proteins (C1q, C2, and C3) and cytokines (IFN-c, TNF-a, and IL-6). Destruction of these immune system proteins by bacterial proteases is a way for the bacteria to suppress both the innate and adaptive immune response [14].

A *dsbA* mutation can also impair certain lipolytic enzymes [14]. Fat-burning enzymes are secreted by the T2SS. These lipases break down lipids to produce free fatty acids and glycerol as by-products [14]. *P. aeruginosa* produces a large amount of LipA, which must first be activated by the chaperon lif [18, 19]. It is encoded in the lipA/lipH operon, along with its cognate foldase LipH, which is also necessary for the production of LipC [19].

5. The regulation of ECF sigma factor AlgU in *P. aeruginosa*

The ECF sigma factor family is a subset of the sigma 70 families of sigma factors. ECF stands for extracytoplasmic factor. These sigma factors are typically sequestered in an inner membrane complex (A complex of proteins encoded by genes in an operon encoding the ECF sigma factor) with the sigma factor residing in the cytoplasmic side of the bacterial cell. In response to different extra cytoplasmic signals or stress signals, the inner membrane complex is cleaved through selective proteolysis of different components and the sigma factor is released for gene expression. This type of regulation is well characterized for *E. coli* and now for *P. aeruginosa* as well (**Figure 4**). The operon encoding the ECF sigma factor typically has a similar organization with conservation shown in the gene arrangement, mode of activation, expression, protein structure (Sigma factor), and consensus elements in the promoters recognized by ECF sigma factors. There is also a likelihood of complementarity of function. For instance, the *E. coli* sigma factor RpoE can complement an AlgT/U mutation. This complementarity of function may exist for other bacterial species.

Typically the AlgT/U sigma factor is sequestered in the cytoplasmic side in association with MucA, which is stabilized by MucB-via interaction with MucA periplasmic domain or C terminus end. In the *mucA22* allele, the encoded MucA is prematurely truncated so the C terminal part that interacts with MucB is not made. The MucB interacting domain is missing. Consequently, the MucA22-AlgT interaction is destabilized as well. This is subject to proteolytic activity from AlgW. Further degradation of the cytoplasmic component of MucA is mediated by MucP and ClpXP as well. This process is called regulated inner membrane proteolysis or RIP for short. Once MucA has degraded, the AlgT/U molecule is free to reach the RNA polymerase so that it can be expressing genes that are part of its regulon. The extracytoplasmic

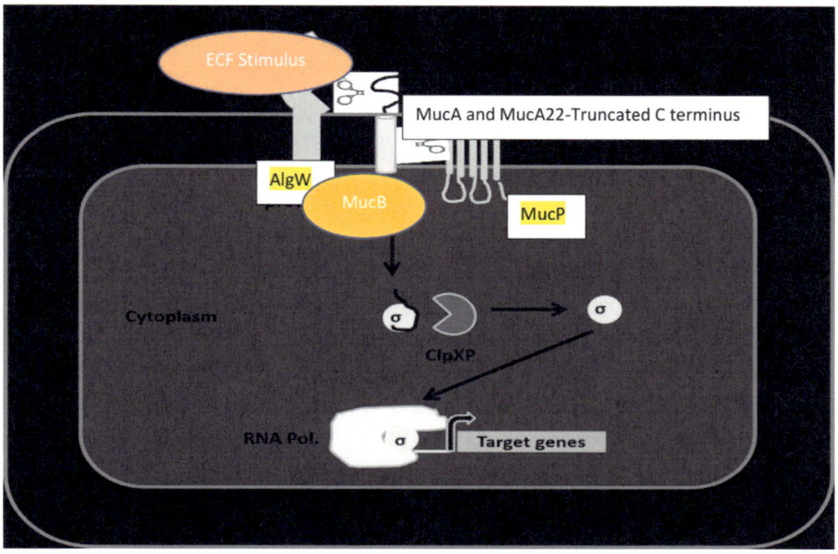

Figure 4.
ECF Sigma factor activation promoter consensus sequence. (a) ECF sigma factor activation and promoter specificity. (b)Truncate MucA22 cannot bind MucB so its attacked by AlgW and degraded releasing Sigma 22 or AlgT. ECf signals are discussed in Text. MucP is another protease indicated in proteolysis. (c) Promoter Sequence Specificity: Different residues and spacer element account for recognition sequence of a consensus promoter.

stimuli that contribute to MucB and MucA degradation are misfolded proteins in the OM or periplasmic space, osmotic stress, antibiotics, and H_zO_2. RIP of MucA leads to activation of σ 22 and expression of the σ 22 regulon. RIP of MucA is accomplished by the proteases AlgW, MucP, ClpXP, and Prc (**Figure 4**) [14, 15, 19].

6. The hypothesis testing: stress response regulon of AlgT/U

As AlgT/U belongs to a highly conserved subfamily of Sigma 70-like proteins called the ECF (Extracytoplasmic Function family of sigma factors) we would anticipate that its regulon contains genes that have been seen to be regulated by other ECF sigma factors in different bacterial species. For instance, we have RpoE in *E. coli*. It may contain genes within its regulon that exist *in P. aeruginosa* and consequently could be part of AlgU/T regulon or any of the other 19 ECF sigma factors found in *P. aeruginosa*. Since AlgT is important for infection, which involves a stress response. Induction of AlgT on infection may likely turn on unique genes that could be important for virulence. Even though molecular genetics for *P. aeruginosa* was well developed with several genetic tools, the genome for *P. aeruginosa* PAO1 was near completion with large sequence contigs available (In 2000!!). Dr. Ohman's group developed several tools to assess AlgT gene function in different strain backgrounds including the *mucA22* allele background (where AlgT is fully active with a copious amount of alginate production.). We resorted to trying a reverse genetic approach to identify targets from AlgT/U. We compared whole cell lysates of PAO1—the prototypic *P. aeruginosa* strain with PDO300 that carries the *mucA22* allele in PAO1 strain. Due to the *mucA22* mutation, we should see the enhanced expression of AlgT-driven genes (ADG). Enhanced expression of ADGs should yield some protein spots that exhibit a higher steady-state level (higher staining protein spots, stained with

Coomassie blue, that binds protein in a most linear fashion) of protein in the *mucA*22 strain PDO300 than PAO1. A visual and densitometric comparison identified several spots that exhibited increased expression in PDO300 than PAO1. This observation told us that the proteomic strategy was working. We also identified proteins involved in alginate biosynthesis to be elevated in the PDO300 strain (*mucA22* allele, alginate overproducing). We chose to use Coomassie over the silver stain as all literature suggested that Coomassie gave more consistent results over decreased sensitivity to silver stain and better results with Edman degradation protocols. Consequently, we consistently identified about 900 proteins that represent about 1/3 of the entire expressed genome of P. aeruginosa (assuming 3000 genes are expressed at any given time). This was the first proteome study of AlgT regulon at that time. Subsequently, other comparisons might have been done. We reliably identified DsbA, Porin F, AlgD, and AlgA in our proteome studies to be elevated in *mucA*22 allele-bearing strain PDO300. This correlates perfectly with the genetic studies back then. Another unidentified protein about 14 KdA size, is YgaU (*E. coli*) ortholog called LysM in *P. aeruginosa* is also elevated in PDO300 (*mucA22*) ([20], unpublished data). We constructed transcriptional lacZ reporter fusions and introduced them into PAO1 and PDO300 strain backgrounds and showed elevated transcription in PDO300 over PAO1. Again, supporting the genetic studies and proteomic results. We also discovered some proteins that exhibited increased accumulation in the non-mucoid state N2-succinylornithine 5-aminotransferase or AruC is one such example [Data not shown 5, [21]]. Porin F, one of the most abundant porins, showed enhanced accumulation in mucoid scenario, but its promoter fusion showed no effective enhancement under mucoid conditions (AlgT+) [6]. Recently sequencing of PA14 was completed [22]. PA14 is more virulent than PAO1. More proteomic work and characterization of *dsbA* can be done in this strain [23].

7. Characterization of *dsbA* Gene through Cassette mutagenesis

Our proteomic studies show that a DsbA-like protein is induced under mucoid conditions. We then cloned the gene inserted a nonpolar gentamycin cassette in it and introduced it into the bacterial cell to create a *dsbA* gene on the chromosome that carries the nonpolar gentamycin cassette insertion [6]. We also created a complementing clone that carries the *dsbA* gene *in trans*. We compared PAO1 wild type to its two variants: PAO1 with *dsbA::Gm* cassette insertion (PAO1 *dsbA* insertion mutant) and PAO1 *dsbA* insertion mutant with a complementing *dsbA* gene *in trans*. Since *dsbA* disruption causes a pleiotropic phenotype [18]. We examined these strains for deficits in virulence factor production and activity (elastase, alginate, twitching motility, alkaline phosphatase, protease activity {Skim milk agar} and sensitivity to DTT). There are two *dsbA*s in *P. aeruginosa* and we characterized pDsbA 1. So far, no function has been ascribed to pDsbA-2. There are two *dsbB*-1 and *dsbB*-2, both of which can be used to regenerate DsbA-1. Thus in our analysis, we showed convincingly that a *dsbA* mutation is deleterious to the production, and secretion of several virulence factors, such as the metalloprotease Elastase. Elastase clearly showed a deficiency in secretion and activity (measured extracellularly). The twitching motility phenotype was reduced [6]. We did not look at individual components of the pilus but the *pilA* gene encodes cysteines and the PilA protein is involved in surface and host cell adhesion, colonization, biofilm maturation, virulence, and twitching, a form of surface-associated motility facilitated by cycles of extension, adhesion, and retraction of T4P fibers [19]. It is also involved in type two secretion and calcium signaling. There is a conserved C terminal disulfide loop (DSL) that when mutated by replacing the "C" residue abrogates assembly of Type 4 pilus or T4P entirely, other substitutions in and around the DSL also affect pilus assembly as measured by twitching motility and structure-based studies and surface assembly of the Type 4 pilus [24]. Suffice it to say, we conclusively showed that we had characterized the right gene, in particular, because two proteins PDsbA-1 and PDsbA-2 encoded by the corresponding genes exist. We only characterized pDsbA-1 because firstly only that gene exhibited enhanced protein accumulation under mucoid scenario. Second, it was at that time beyond the scope of the Author's PhD. work. We then focussed on the transcription of the *dsbA* gene: We showed through promoter fusions that *dsbA* gene expression was enhanced three-fold under mucoid conditions. We then focussed on the promoter region of the *dsbA* gene. The *Pseudomonas* genome was partially sequenced and from the various contigs, we could PCR amplify the promoter region to generate transcriptional fusions to *lacZ* gene in an integrative vector pMS7. Single copy gene fusions were created in PDO300 (*mucA22* allele, Mucoid phenotype, AlgT+ scenario) and PAO1 (the prototypic *P. aeruginosa* strain which is nonmucoid and AlgT). The genomic organization of the single-copy *lacZ* fusions was verified by genomic PCR. Subsequently, the author characterized DsbA-like proteins in *H. influenzae* (por periplasmic oxidoreductase) and in the plant symbiont *S. meliloti* (TlpA: thioredoxin-like protein A.) Por also exhibited high sensitivity to DTT. We discuss more about Por and TlpA in the next sections [25, 26]. The disruption studies i.e., *dsbA* disruption by

Gene name	Bacterial Species	Phenotype	Virulence defect	Comments
dsbA	*E. coli* **0157**	Reduced production of virulence factors	Reduced colonization of *C. elegans* gut and biofilm formation	Reduced virulence phenotypeAt 9 days >50% better survival of *C. elegans* mutant strain.
dsbA	*Pseudomonas aeruginosa*	Reduced Twitching motility, *pilA* expression reduced	Reduced slow and fast killing in the *C. elegans* assay.	DsbA alters several virulence factors in the bacterium.
dsbA	*Erwina carotovora*	Less secreted enzymes, production of quorum sensing signal	Motility is highly reduced in *dsbA* mutant. Rotted tissue in potato tuber assay was much reduced in the *dsbA* mutant.	DsbA is responsible for so many traits.Secretome and other things.Secretion of quorum sensing signals.
sdbA	*Streptococcus pyogenes*	Catalyst for the enzyme SpeA	SpeA is oxidized by SdbA and maintains biological function	In the crystal structure of SpeA, Cys87 and Cys98 are linked by disulfide bond

A short list of 4 DsbA orthologs from different bacterial species all having a common theme- defective in the production of virulence factors. This invariably results in a reduction in virulence of the associated pathogen harboring a mutation in the respective DsbA.

Table 1.
DsbA orthologs and their role in different bacteria.

cassette mutagenesis, exhibits loss of virulence factor maturation as shown in other bacterial species (**Table 1**).

8. A close look at the *dsbA* gene/operon: primer extensions

Since the transcriptional fusion data suggested a clear induction of *dsbA* gene under mucoid conditions, we chose to examine the putative promoter region around the *dsbA* open reading frame in greater detail. We used a simple primer extension analysis procedure on whole cell mRNA isolated from PAO1, PDO300 and PDO300 with *dsbA in trans* [27]. As the data clearly shows in **Figure 5**, right panel, appearance of a novel transcript (T1) under mucoid conditions. It also appears to be driven by promoter elements that are the consensus for a ECF sigma factor (left panel, **Figure 5**) [28]. This data is also replicated in the Clinical isolate FRD1 and its derivative FRD440 [29]. Now we have a very substantial study to show enhanced expression of the *dsbA* gene, accumulation of the protein DsbA (Proteomic data) under mucoid conditions (AlgT+ scenario).

Considering this study, there have been numerous characterizations of DsbA and DsbA-like proteins in several bacterial systems [6]. We have a wealth of data suggesting the importance of these protein systems (including DsbB) in virulence determination and maturation of virulence factors [30].

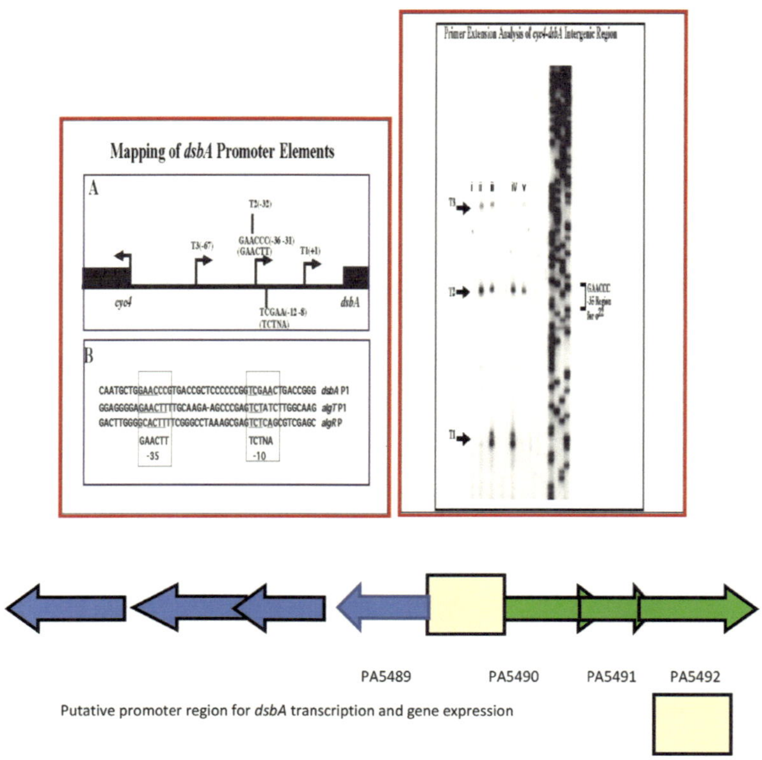

Figure 5.
Analysis of dsbA *promoter by primer extension. Top Left panel A Arrangement of the transcript initiation site. Top Right Panel Primer extension reaction Lane Order No RNA PAO1, PDO300, FRD1 & FRD440. Three initiation sites with T1 appearing in mucoid conditions. Bottom Panel arrangement of orf's schematic diagram. Top Left Panel in B we see the alignment of the promoter sequence with the consensus −35 and −10 regions with a 15/17 nucleotide spacer region.*

9. Novel regulation of the *dsbA* gene in *P. aeruginosa*: comparison with *E. coli* and others

Interestingly our extensive work with *dsbA* discovered this novel mechanism of regulation. So far my literature searches have identified only one-gram-positive organism where this is replicated: direct control of expression of *dsbA* gene by the ECF sigma factor. In our case, we have proteomic gene expression studies {promoter fusion studies in various strains} and primer extension studies that show direct evidence of AlgT/U directed transcription of the *dsbA* gene [5 and Primer extension data]. This is clearly distinct from *E. coli* where ECF factor RpoE does not directly enhance *dsbA* {EC} gene expression. In *E. coli* the CpxA/R two-component regulator directs transcription of the *dsbA* gene along with other genes that encode proteins involved in folding and assembly of proteins. The first characterized genes in the CpxA/R regulon were Peptidyl prolyl isomerase, DegP protease, DsbA, among others [31, 32]. The regulon is much larger now. SigL is an ECF sigma factor identified in *Mycobacterium tuberculosis*, disruption of the sigL gene causes a loss in virulence indicated by increased survival time in the BALB/c mice. Mpt53, annotated as a DsbE-like protein, is a gene in the SigL regulon. It has been characterized biochemically and structurally [33]. It contains a thioredoxin active site and is a strong oxidant, in contrast to the weak reducing activity of *E. coli* DsbE. These biochemical data suggest a possible role for Mpt53 as an extracellular oxidant that may be required for proper folding of reduced unfolded secreted proteins, a function more similar to that of *E. coli* DsbA [33]. It seems the *P. aeruginosa* DsbA-1 protein's gene is regulated in a manner similar to Mpt53: Directly via an ECF sigma factor [33].

10. DsbA in biofilms: gene expression and disruption studies

DsbA has been implicated in Biofilm formation as well [12]. Inés Reigada *et al* a dual mix species biofilm of *P. aeruginosa* and *S. aureus,* proteins were assessed in the surfaceome and exoproteome compartments were determined for both bacterial species and compared between monospecies biofilms and dual-species biofilms [12]. DsbA levels were elevated in *P. aeruginosa* surfaceome. Alginate is probably not critical for biofilm formation [34]. So far our tests with the PAO1 *dsbA*::Gm strain suggest the *dsbA* disruption augments the biofilm kinetics. We see slightly faster biofilm formation in the *dsbA* mutant [Preliminary findings]. Quite similarly a *dsbA* mutant in *Pseudomonas putida* also shows enhanced exopolysaccharide production and increased biofilm formation on plastic [35, 36]. However, disruption studies in other organisms show a reduction in biofilm formation [30]. I believe that in *P. aeruginosa dsbA* contribution is essential along with alginate in biofilm mode. Especially in Cystic fibrosis [34], where the presence of alginate and excess of DsbA (due to induction) may be helpful in the production of other virulence determinants [15, 18]. Regulation of biofilm through the biofilm master regulator CsgD and fimbrial subunit CsgA. Deletion of either of these proteins causes a lack of enhancement of biofilm formation seen due to disruption of *dsbA* in *P. aeruginosa* [37]. Now with the advent of microarrays and such technologies, we can assess *dsbA* expression in biofilm and sessile bacterial populations as well. Mark A Webber's Group showed, using the transposon mutagenesis approach, a strong temporal contribution to biofilm fitness for some genes, including somewhere expression changed between being beneficial or detrimental depending on the stage at which they are expressed, including *dksA* and *dsbA*. These studies were carried out on *E. coli* [38]. DsbA is involved in early biofilm for *E. coli* [38].

11. Future potential of DsbA research and avenues

Interestingly DsbA and DsbA-like proteins are not limited to gram-positive and negative bacteria like *P. aeruginosa* and *E. coli* alone. But in Eukaryotes specific advances have been made. Briefly, in eukaryotes, protein disulfide isomerases primarily facilitate the formation of protein disulfide bonds between cysteine residues. This process takes place in the endoplasmic reticulum lumen environment. Biology of PDI's or protein disulfide isomerase-like molecules from five different species of malaria parasites—*Plasmodium berghei*, *Plasmodium yoeli*, *Plasmodium knowlesi*, and *Plasmodium falciparum*—are potential targets for research. Based on their chromosomal positions, *P. falciparum* four investigated protein disulfide isomerases have been given the names PfPDI-8, PfPDI-9, PfPDI-11, and PfPDI-14. The prototype PDI molecule with two thioredoxin domains (including CGHC active sites) and a C-terminal Endoplasmic reticulum retrieval signal, SEEL, is PfPDI-8. Recently, Fiona Angrisano *et al* demonstrate that a plasmodium protein (PDI-Trans/PBANKA_0820300) is a PDI that is male specifc, surface-expressed, and is essential for fertilization/transmission. It exhibits disulfide isomerase activity that is up-regulated postgamete activation. Using bacitracin-a PDI inhibitor they show a reduction in male gamete activation and fusion. This PDI activity can be inhibited by an antibody directed against the extracellular domain [39, 40].

With respect to DsbA and bacterial pathogenesis lot of research has been done. First and foremost many downstream targets of DsbA have been identified and the contribution of DsbA to pathogenesis has been assessed by demonstrating that disruption of the gene for DsbA has a deleterious effect on pathogenesis due to reduction in the expression of virulence factors most notably as functional virulence molecules at the correct locale in the cell [41, 42].

The focus areas are likely to be the following:

- Design of broad-spectrum mimetics that can bind to the hydrophobic domain of the DsbA catalytic site [31]. This has been shown to disrupt the DsbA function. Assessment of in vivo function of these mimetics.

- Identification of downstream targets both in biofilm and planktonic versions of a given bacterial species. Special focus on inner membrane and outer membrane proteins. Potential proteins involved in Quorum sensing, etc. [42].

- Assessment of thiol status of proteins and correlation with protein amounts, so a thiol to protein quantity determination as an assessment to a protein's ability to be functional. Since a protein that is likely to have disulfide bonds will likely be more stable, so a protein that loses its disulfide bonds becomes less stable and consequently it steady-state levels drop and it is also less likely to be functional. Thus a correlation of thiol status of given major virulence factors or complexes to their amount is a key finding for a targeted attack on novel virulence determinants in newly emerging pathogenic bacterial species. Some assessments are being done for *in vivo* thiol status [43].

- A combination of metabolomic and proteomic and genomic approaches to obtain a systemic understanding of DsbA and downstream protein targets in bacterial pathogenesis [42]. Scientists have developed metabolomic profiles for bacteria that might reside in the CF lung: For *P. aeruginosa* it was Cyanide production [44]. Later studies showed that beyond identifying the presence of a specific molecule, groups of *P. aeruginosa*-colonized CF patients could be differentiated from noncolonized CF patients based on overall volatile breath profiles [45].

- Finally, increased understanding is likely to generate avenues for clinical trials incorporating targeted therapy on DsbA and DsbA targets at least in biofilms. Multiple strategies are used to limit biofilm-based infections on implants [46, 47]. Research on DsbA mimetics and other targets could yield new treatment modalities [46, 48, 49].

12. Summary

In summary, we have examined my old work and how we demonstrated the direct involvement of AlgT with *dsbA* gene expression. This is unique to *P. aeruginosa*. So far literature searches have not identified a similar regulatory mechanism in any other gram-negative bacterium. We have recently seen a similar ECF sigma factor directed control on a protein disulfide isomerase gene in a gram-positive bacterium. Our hypothesis as to why the DsbA gene is upregulated in a mucoid scenario is the fact that I think the alginate exopolysaccharide creates a difficult capsule to allow other virulence factors to go out of the cell. I have shown using western blot (data not shown) that Elastase secretion on a per cell basis (normalized for OD_{600nm} is lower in cells producing alginate versus cells producing no alginate. Intracellular accumulation of elastase was identical in both alginate-producing and nonproducing strains. To support this hypothesis we should examine situations or bacterial systems here such induction of *dsbA* is seen routinely. We can then evaluate *dsbA* function towards pathogenesis by looking at steady state levels of virulence determinants and components of virulence factor secretion machinery. In addition, since following *dsbA* gene induction and concomitant accumulation of *dsbA* protein we could see a better disulfide bond status—so maybe compare the protein levels to thiol status in proteins [2, 43, 50, 51]. In addition, the ECF response to induce *dsbA* gene and subsequently DsbA protein levels could be a molecular switch to up the thiol status of the *Pseudomonas* bacterial cell in response to harsh conditions from host immune response. This was seen with acidophilic iron-oxidizing bacterium *Leptospirillum ferriphilum* where exposure to ROS upped the activity of intracellular DsbA-like proteins [52]. Along with metabolomic profiling of compounds can be done. It is clear from various studies that disulfide bonds are crucial for protein stability and that DsbA facilitating disulfide bonds could be a crucial target for future mimetics that may be targeted to the hydrophobic domain of its enzymatic region [48]. Alternatively, novel downstream targets that affect pathogenesis, even more, could be targeted for drug design or mimetics. Very briefly I discovered more disulfide isomerase-like proteins *por* in *H. influenzae* and TlpA in *S. meliloti*. A TlpA mutant caused a complete loss in nodule forming capacity in the mutant strain. That could be recovered in the complementation clone (Data not shown). With Por (Por: *P*eriplasmic *o*xido*r*eductase). I characterized the periplasmic fraction and in a small pilot, experiment identified a novel target RibH a component of the Riboflavin synthase (data not shown. See Figure 3 for MSA alignments of DsbA orthologs from different species of bacteria).

13. Conclusion

Briefly, this chapter describes the involvement of DsbA in the pathogenesis of *P. aeruginosa*. We describe *dsbA* gene characterization with gentamycin cassette mutagenesis. The effect of disruption of the *dsbA* gene on virulence factor production is provided. More recently on biofilm formation as well. Since *P. aeruginosa* is an important pathogen we are constantly developing strategies to identify ways to eliminate it. A series of novel spirothiazolopyridine derivatives were designed and prepared with strong antimicrobial and antifungal properties [53]. Such compounds are likely to enhance our ability to stay abreast of AMR and novel pathogens. In addition, novel diagnosis methods being developed should be used to assess microbial contamination through innovative methods of diagnoses [54]. This collection of studies of Dsb proteins from different bacteria clearly shows their strong involvement in bacterial pathogenesis. We propose a more defined focus on the studies of these proteins and the development of methods to suppress their function. We also show a direct interaction of AlgT with the *dsbA* promoter a so far unique regulation not seen in gram-negative bacteria. Since Dsb proteins might be involved in biofilm we suggest a focus on their role in biofilm. We also suggest study of phenazine compounds like pyocyanin due to their role in bacterial pathogenesis [55]. It would be fascinating to see the involvement of DsbA in phenazine compound secretion and their redox status.

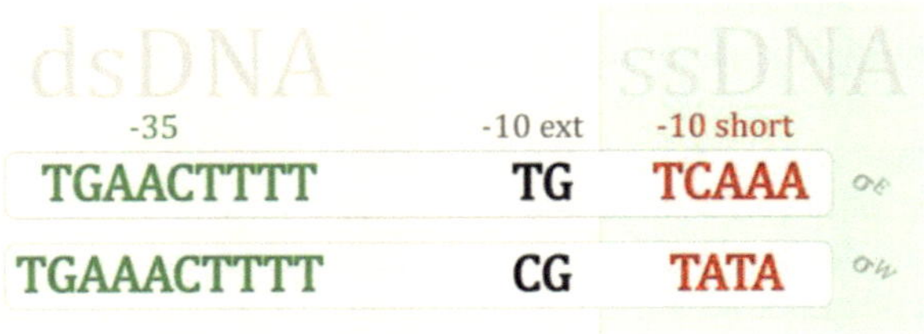

Acknowledgements

I would like to thank my Ph.D. Mentor Dr. Dennis E. Ohman, Chair Department of Microbiology and Immunology, Virginia Commonwealth University, Richmond, VA, USA. He gave all the care and encouragement for the development of this stress response regulon hypothesis and the use of proteomics to identify components of this pathway. A similar proteomic strategy was used by other members of the Dennis Ohman Laboratory who contributed to my work through helpful discussions. I am thankful to Dr.(s) Sang Jin Suh, Laura Suh, Lynn Wood, Sumita Jain, Kerain Grande and Joanne Lee Johnston. Further thanks to members of my family, parents and brother who supported me through tough times of my Ph.D. research work. A special thanks to close friends like Dr. Gobardhan Das and Dr. Arshad Jilani. A special thanks to join me on this book chapter n to Dr. Sunil Kumar contributed to the section on Eukaryotic PDIs. His insight into their biology is appreciated.

Author details

Sunil Kumar[1] and Sonal Malhotra[2*†]

1 Department of Medical Lab Technology, Chandigarh University, Gharuan Kharar, Mohali, Punjab, India

2 Infectious Diseases (Hybrid Mode), James Lind Institute, Geneve, Switzerland

*Address all correspondence to: sonalmalhotra07@gmail.com

† A Swiss EduQua certified private education Institute.

References

[1] GuIlhot C et al. Evidence that the pathway of disulfide bond formation in *Escherichia coli* involves interactions between the cysteines of DsbB and DsbA. Proceedings of the National Academy of Sciences of the United States of America. 1995;**92**:9895-9899

[2] Radzinski M, Oppenheim T, Metanis N, Reichmann D. The cys sense: Thiol redox switches mediate life cycles of cellular proteins. Biomolecules. 2021;**11**:469

[3] Poole LB. The basics of Thiols and Cysteines in redox biology and chemistry. Free Radical Biology and Medicine. 2015;**1**:148-157

[4] Reardon-Robinson ME, Ton-That H. Disulfide-bond-forming pathways in Gram-positive bacteria. Journal of Bacteriology. 2016;**198**:746-754

[5] Lee SF, Li L, Jalal N, Halperin SA. Identification of a thiol-disulfide oxidoreductase (SdbA) catalyzing disulfide bond formation in the superantigen SpeA in *Streptococcus pyogenes*. Journal of Bacteriology. 2021;**203**:00153-00121

[6] Malhotra S et al. Proteome analysis of the effect of mucoid conversion on global protein expression in pseudomonas aeruginosa strain PAO1 shows induction of the disulfide bond isomerase, dsbA. Journal of Bacteriology. 2000;**182**:6999-7006

[7] Santos-Martin C et al. Structural bioinformatic analysis of DsbA proteins and their pathogenicity associated substrates. Computational and Structural Biotechnology Journal. 2021;**19**:4725-4737

[8] David M, De Oliveira P, et al. Antimicrobial resistance in ESKAPE pathogens review. Clinical Microbiology Reviews. **33**:181-219

[9] Stover CK et al. Complete genome sequence of *Pseudomonas aeruginosa* PAO1, an opportunistic pathogen. Nature. 2000;**406**(6799):959-964

[10] Lonetto MA et al. Discovery of extra cytoplasmic functions factors. Molecular Microbiology. 2019;**112**(2):348-355

[11] May TB et al. Alginate synthesis by *Pseudomonas aeruginosa*: A key pathogenic factor in chronic pulmonary infections of cystic fibrosis patients. Clinical Microbiology Reviews. 1991;**4**:191-206

[12] Reigada I et al. Surfaceome and exoproteome dynamics in dual-species *Pseudomonas aeruginosa* and *Staphylococcus aureus* biofilms. Frontiers in Microbiology. 2021;**12**:1-23. Article 672975

[13] Franklin MJ et al. Biosynthesis of the *Pseudomonas aeruginosa* extracellular polysaccharides, alginate, Pel, and Psl. Frontiers in Microbiology. 2011;**2**:1-16. Article #137

[14] Moradali F et al. *Pseudomonas aeruginosa* lifestyle: A paradigm for adaptation, survival, and persistence. Frontiers in Cellular and Infection Microbiology. 2017:1-29. Article #39

[15] Heath Damron F, Goldberg JB. Proteolytic regulation of alginate overproduction in *Pseudomonas aeruginosa*. Molecular Microbiology. 2012;**84**(4):595-607

[16] Ramsey DM, Wozniak DJ. Understanding the control of *Pseudomonas aeruginosa* Alginate synthesis and the prospects for management of chronic

infections in cystic fibrosis. Molecular Microbiology. 2005;**569**(2):309-322

[17] Braun P et al. Maturation of *Pseudomonas* Elastase. Journal of Biological Chemistry. 2001;**276**(28):26030-26035

[18] Ha U-H et al. DsbA of pseudomonas aeruginosa is essential for multiple virulence factors. Infection and Immunity. 2003;**71**:3, 1590-1595

[19] Jurado-Martín I et al. Pseudomonas aeruginosa: An audacious pathogen with an adaptable arsenal of virulence factors. International Journal of Molecular Sciences. 2021;**22**:3128

[20] Raivio TL. Everything old is new again: An update on current research on the Cpx envelope stress response. Biochimica et Biophysica Acta. 2014;**1843**:1529-1541

[21] Itoh Y. Cloning and characterization of the aru genes encoding enzymes of the catabolic arginine succinyltransferase pathway in Pseudomonas aeruginosa. Journal of Bacteriology. 1997;**179**(23):7280-7290

[22] Lee DG, Urbach JM, Wu G, et al. Genomic analysis reveals that *Pseudomonas aeruginosa* virulence is combinatorial. Genome Biology. 2006;7:R90

[23] Wareham DW et al. The *Pseudomonas aeruginosa* PA14 type III secretion system is expressed but not essential to virulence in the *Caenorhabditis elegans–P. aeruginos*a pathogenicity model. FEMS Microbiology Letters. 2005;**242**:209-216

[24] Harvey H et al. Single-residue changes in the C-terminal disulfide-bonded loop of the pseudomonas aeruginosa type IV pilin influence pilus assembly and twitching motility. Journal of Bacteriology. 2009;**191**(21):6513-6524

[25] Loferer H et al. *Bradyrhizobium japonicum* TIpA, a novel membrane anchored thioredoxin-like protein involved in the biogenesis of cytochrome aa3 and development of symbiosis. The EMBO Journal. 1993;**12**(9):3373-3383

[26] Tomb JF. A periplasmic protein disulfide oxidoreductase is required for transformation of *Haemophilus influenzae* Rd. Proceedings of the National Academy of Sciences of the United States of America. 1992;**89**(21):10252-10256

[27] Smith DR. Primer extension. Methods in Molecular Biology. 1993;**18**:373-378

[28] Guzina J, Djordjevic M. Mix-and-matching as a promoter recognition mechanism by ECF σ factors. BMC Evolutionary Biology. 2017;**17, (Suppl. 1)**:12

[29] Min KB et al. Nonmucoid conversion of mucoid *Pseudomonas aeruginosa* induced by sulfate-stimulated growth. FEMS Microbiology Letters. 2014;**360**(2):157-166

[30] Lee Y, Sejong O, P W. Inactivation of the Pseudomonas *putida* KT2440 *dsbA* gene promotes extracellular matrix production and biofilm formation. FEMS Microbiology Letters. 2009;**297**(1):38-48

[31] Totsika M et al. Inhibition of diverse dsba enzymes in multi-dsba encoding pathogens. Antioxidants & Redox Signaling. 2018;**29**(7):653-666

[32] Tran NT et al. Defining the regulon of genes controlled by σE, a key regulator of the cell envelope stress response in *Streptomyces coelicolor*. Molecular Microbiology. 2019;**112**(2):461-481

[33] Goulding CW et al. Gram-positive dsbe proteins function differently from gram-negative dsbe homologs. A structure to function analysis of dsbe from *Mycobacterium tuberculosis*. Journal of Biological Chemistry;**279**:3516-3524

[34] Kiedrowski MR, Bomberger JM. Viral-bacterial co-infections in the cystic fibrosis respiratory tract. Frontiers in Immunology. 2018;**9**:1-12. Article #3067

[35] Cystic Fibrosis Foundation. 2015 CFF Patient Registry Annual Data Report. 2016. pp. 1-94

[36] Kay E et al. Two GacA-dependent small RNAs modulate the quorum-sensing response in *Pseudomonas aeruginosa*. Journal of Bacteriology. 2006;**188**:6026-6033

[37] Laverty G, Gorman SP, Gilmore BF. Biomolecular Mechanisms of *Pseudomonas aeruginosa* and *Escherichia coli* Biofilm Formation. Pathogens. 2014;**3**:596-632

[38] Holden ER et al. Massively parallel transposon mutagenesis identifies temporally essential genes for biofilm formation in *Escherichia coli*. Microbial Genomics. 2021;**7**:000673

[39] Angrisano F et al. Male-specifc protein disulphide isomerase function is essential for plasmodium transmission and a vulnerable target for intervention. Nature Scientific Reports. 2019;**9**:1830

[40] Mahajan B et al. Protein disulfide isomerase assisted protein folding in malaria parasites. International Journal for Parasitology. 2006;**36**(9):1037-1048

[41] Łasica AM, Jagusztyn-Krynicka EK. The role of Dsb proteins of gram-negative bacteria in the process of pathogenesis. FEMS Microbiology Reviews. 2007;**31**(5):626-636

[42] Arts IS et al. Dissecting the machinery that introduces disulfide bonds in *Pseudomonas aeruginosa*. mBio. 2013;**4**(6):00912-00913

[43] Leichert LI, Jakob U. Protein thiol modifications visualized in vivo. PLoS Biology. 2004;**2**(11):e333

[44] Gilchrist FJ et al. Exhaled breath hydrogen cyanide as a marker of early *Pseudomonas aeruginosa* infection in children with cystic fibrosis. ERJ Open Research. 2015;**1**:00044-02015

[45] Purcaro G et al. Volatile fingerprinting of *Pseudomonas aeruginosa* and respiratory syncytial virus infection in an in vitro cystic fibrosis co-infection model. Journal of Breath Research. 2018;**12**:046001

[46] Van Dyck K, Pinto RM, Pully D, Van Dijck P. Microbial Interkingdom Biofilms and the Quest for Novel Therapeutic Strategies. Microorganisms. 2021;**9**:412

[47] Verderosa AD, Totsika M, Fairfull-Smith KE. Bacterial biofilm eradication agents: A current review. Frontiers in Chemistry. 2019;**7**:1-17. Article 824

[48] Martínez OF, Duque HM, Franco OL. Peptidomimetics as Potential Anti-Virulence Drugs Against Resistant Bacterial Pathogens. Frontiers in Microbiology. 2022;**13**:831037

[49] Pandeya A et al. Periplasmic targets for the development of effective antimicrobials against gram-negative bacteria. ACS Infectious Diseases. 2020;**6**(9):2337-2354

[50] Hillion M, Antelmann H. Thiol-based redox switches in prokaryotes. Biological Chemistry. 2015;**396**(5):415-444

[51] Sutton TR et al. A robust and versatile mass spectrometry platform for comprehensive assessment of the thiol redox metabolome. Redox Biology. 2018;**16**:359-380

[52] Norambuena J, Flores R, Ca'rdenas JP, Quatrini R, Cha'vez R, et al. Thiol/disulfide system plays a crucial role in redox protection in the acidophilic

iron-oxidizing bacterium *Leptospirillum ferriphilum*. PLoS One. 2012;7(9):e44576

[53] El-Sofany WI et al. Boosting the antimicrobial performance based on new fused spirothiazolidine framework analogs. Journal of the Iranian Chemical Society. 2022;**19**:4223-4236

[54] Ahmed BS et al. Development of specific nano-antibody for application in selective and rapid environmental diagnoses of *Salmonella arizonae*. Biointerface Research in Applied Chemistry. 2020;**10**(6):7198-7208

[55] Barakat KM et al. Production and characterization of bioactive pyocyanin pigment by marine *Pseudomonas aeruginosa* OSh1. Research Journal of Pharmaceutical, Biological and Chemical Sciences. 2015;**6**:933-943

Chapter 2

Searching for the Resistance Interactome of *Pseudomonas aeruginosa*

Diana Carolina Castaño, Juan David Patiño-Salazar and Mauricio Corredor

Abstract

Pseudomonas aeruginosa is one of the most important emerging Gram-negative pathogenic bacilli worldwide. The development of antibiotic resistance and its ability to adapt to multiple environmental conditions keep triggering alarms in global hospitals since the invasion of different types of tissues. This facultative anaerobe can adapt easily to aerobic or anaerobic conditions. It invades tissues, such as the lung, gastrointestinal tract, skin, renal system, and urinary tract, to the extreme of causing a variety of punctate gangrene. The considerable size of its genome (core and accessory genome) shows that this bacterium carries a huge battery of genes that allow it to develop resistance to various antibiotics, emerging as an MDR bacterium. The most studied mechanisms for resistance development have been *quorum sensing* and biofilm formation, among others. The research of resistance genes has been a long and time-consuming task. Genes such as CARB-3, CARB-4, PSE-1 (CARB-2), PSE-4 (CARB-1), OXA-18, OXA-2, OXA-21, OXA-10 (PSE-2), GyrA, GyrB, OprM, OprJ, OprN, MexB, MODx, MexF, and MexY, are among the best-characterized genes in *P. aeruginosa*. Another group of not-so-conventional genes is the methyltransferases, which have been negligible studied in *P. aeruginosa*. In this article, we propose to give a state of the art of the most important resistance genes of *P. aeruginosa* and their relationship with the interactome-resistome.

Keywords: *Pseudomonas aeruginosa*, interactome, resistome, antibiotic resistance genes, pangenomics

1. Introduction

In the past years, antibiotic resistance of *Pseudomonas aeruginosa* has been reported as an emerging global health problem by the WHO [1]. P. aeruginosa is one of the most prevalent nosocomial pathogens, responsible for 57% of overall hospital-acquired infections because it is a ubiquitous pathogen [2]. In the twentieth century, this bacterium was not a global health problem and, in many hospitals worldwide, was treated locally with aminoglycosides. Although, antibiotic resistance made it necessary to replace them with b-lactams, fluoroquinolones, and polymyxins. *P. aeruginosa* causes infections in the blood as septicemia, in the lungs as pneumonia for respiratory complications, in cystic fibrosis patients, and in the skin as epidermolysis bullosa or folliculitis [3, 4]. It can contaminate any part of the body after surgery. Now, it needs to be studied more seriously, precisely because the treatment with antibiotics has not only developed rapidly antibiotic-resistant bacteria (ARB), since it has a rapid dispersion, and it becomes multidrug-resistant (MDR) [5]. Resistant strains spread from one person to another through the infection of the mouth, lips, hair, saliva, hands, skin, and nails, likewise by infected apparatus or surfaces, but also through wind, water, and soil [6], yield a rapid spread and the development of multidrug resistant (MDR) [5], **Figure 1**.

Genomics emerges as a fundamental tool to give us a new hand to combat ARB [7]. Nonetheless, to better know the complex gene interactions, the genome size of P. aeruginosa complicates the task of identifying different genes responsible for specific antibiotic resistance. This approach has been used for years but is nowadays outdated. New strategies such as the omics allowed decoding of sequences, gene organization, and synteny; likewise, as quorum sensing communication [8, 9], epigenetic labeling on DNA and RNA, and resistome interactions [10]. This evidence is a multifaceted problem that cannot continue to be focused as a reductionist manner. Genes and proteins are arranged in an intricate chain of events to simplify the cellular activity in a discrete compartment of a pair of genes. Genes and proteins network could be the strategy of cellular events giving success to living beings, including pathogen bacteria.

The genes and proteins interactions are fundamental elements to breakthrough biological processes. The comprehensive identification of the gene-protein interaction, messenger RNA-protein association, and protein-protein interactions (PPIs) is well-known today as the interactome [11]. The research of antibiotic interactome (resistome) provides a bioinformatics framework to understand the organisms as an integrated system (system biology). In other words, the resistome is the part of the

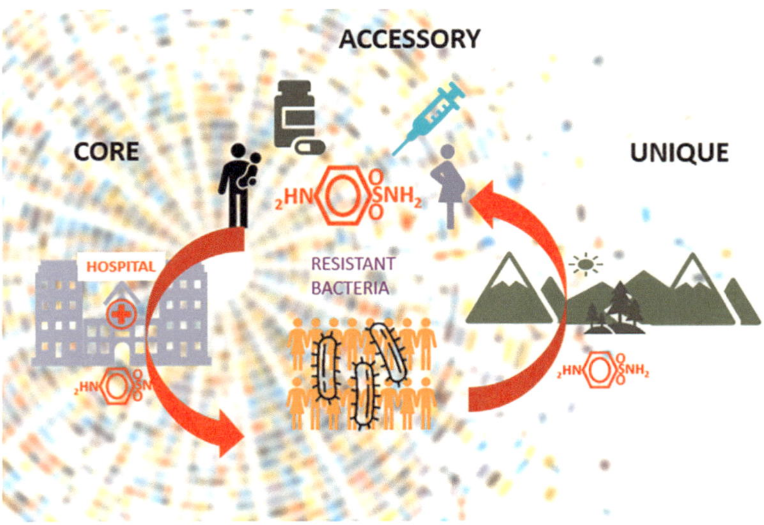

Figure 1.
Resistant genes of antibiotics coming from Pseudomonas aeruginosa *strains (all strains or core genome) and the environment (some strains with accessory and unique genomes) as horizontal genes transfer, HGT.*

interactome that manages antibiotic resistance [12]. Resistome could be part of the core or accessory genome in the whole bacteria species [13–15]. However, it is not enough to study resistome using specific antibiotic resistance genes. Innovative approaches are necessary to understand the tangle of intricate steps of cellular regulation and predicting gene and protein networks.

2. Interactome-resistome

Computational advances allow the deep study of protein-protein interactions (PPIs), but this is not possible for all species. Some species have been more observed than others, as in the case of *P. aeruginosa,* given the clinical interest in recent years. Resistome genes comprise essential and nonessential genes and are part of the accessory or core genome, including unique genes. Some of these genes are gained by horizontal transfer. Most of the genome or the full set of genes is not related to drug resistance. Many resistant genes are for extreme cold or heat, high salt concentration, toxic compounds, UV, or radioactivity exposition, etc. [16]. However, the resistome can include many genes related to different challenges. The network between genes or proteins represents biological data, whereas nodes represent biological entities connected by edges, meaning associations between them, which can be diagrammed and computed [17].

Despite the few works in interactomics, new results are arising. The *P. aeruginosa* protein-protein interaction (PPI) network was developed 10 years ago [18]. This study integrated various traits of genes and proteins at the genome level using a machine learning approach to predict a genome-scale PPI network of *P. aeruginosa*. A total of 54,107 interactions spanning 4,181 proteins were predicted. Using a high-confidence network that combines high-throughput predicted interactions, the author found 3,343 proteins and 19,416 potential interactions.

One of the best summaries of interactomes is PseudomonasNet (http://www.inetbio.org/pseudomonasnet/index.php). As they said: this is a database of genome-wide co-functional networks for *P. aeruginosa*. By integrating nine distinct data types, PseudomonasNet covers more than 98% of the 5,572 coding genes of the PAO1 strain with 203,118 co-functional links [19].

The interactome is not just a protein-protein interaction. For example, the RNA-binding protein Hfq interacts with mRNAs, affecting mRNA translation, either alone or together, with the small regulatory noncoding RNAs, and its degradation in *P. aeruginosa* [15]. The authors find that most of the Hfq interactions are not conserved between the different strains studied, and the strain-specific interactions are due to diverse sets of accessory genes.

3. Antibiotics resistance genes and mechanisms in *P. aeruginosa*

In *P. aeruginosa*, the antibiotic resistance can be classified into five distinct categories according to action mechanisms: inactivation of antibiotics by modifying enzymes, modification of antibiotics target, global response efflux pumps conferring multidrug resistance, antibiotic resistance by alterations in the cell wall charge, and porins modulating reduction of permeability to antibiotics. The genes associated with these mechanisms are listed in **Table 1**.

Gene	Product	Classification
alr	Alanine racemase (EC 5.1.1.1)	antibiotic target in susceptible species
ampC	Class C beta-lactamase (EC 3.5.2.6) => PDC family	antibiotic inactivation enzyme
aph	Aminoglycoside 3'-phosphotransferase (EC 2.7.1.95) => APH(3')-II/APH(3')-XV	aminoglycoside resistance gene; antibiotic inactivation enzyme
arnB	UDP-4-amino-4-deoxy-L-arabinose–oxoglutarate aminotransferase (EC 2.6.1.87)	protein altering cell wall charge conferring antibiotic resistance
arnC	Undecaprenyl-phosphate 4-deoxy-4-formamido-L-arabinose transferase (EC 2.4.2.53)	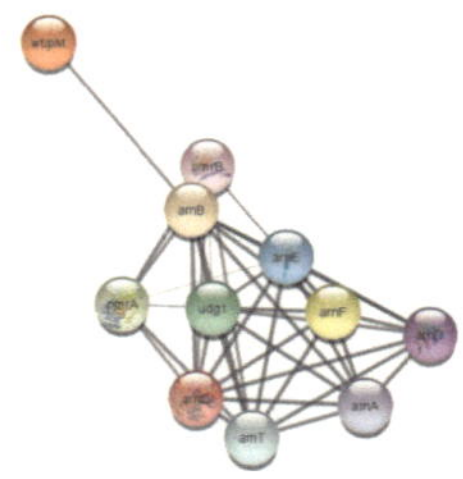
arnD	4-deoxy-4-formamide-L-arabinose-phosphoundecaprenol deformylase ArnD (EC 3.5.1.n3)	
arnE	Undecaprenyl phosphate-aminoarabinose flippase subunit ArnE	
arnF	Undecaprenyl phosphate-aminoarabinose flippase subunit ArnF	
arnT	Undecaprenyl phosphate-alpha-4-amino-4-deoxy-L-arabinose arabinosyl transferase (EC 2.4.2.43) @ Polymyxin resistance protein PmrK	
cat	Chloramphenicol O-acetyltransferase (EC 2.3.1.28) => CatB family	antibiotic inactivation enzyme; chloramphenicol resistance gene
dadX	Alanine racemase (EC 5.1.1.1)	antibiotic target replacement protein
ddl	D-alanine–D-alanine ligase (EC 6.3.2.4)	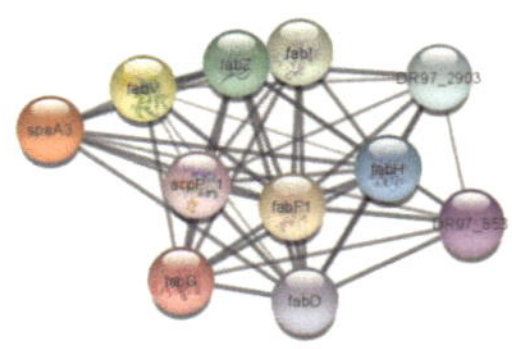
dxr	1-deoxy-D-xylulose 5-phosphate reductoisomerase (EC 1.1.1.267)	
fabF1	3-oxoacyl-[acyl-carrier-protein] synthase, KASII (EC 2.3.1.179)	
fabF2	3-oxoacyl-[acyl-carrier-protein] synthase, KASII (EC 2.3.1.179)	
fabG	3-hydroxyacyl-CoA dehydrogenase (EC 1.1.1.35), FabG4	
fabI	Enoyl-[acyl-carrier-protein] reductase [NADH] (EC 1.3.1.9)	
folA	Dihydrofolate reductase (EC 1.5.1.3)	
folP	Dihydropteroate synthase (EC 2.5.1.15)	
fusA1	Translation elongation factor G	
fusA2	Translation elongation factor G	
gidB	16S rRNA (guanine(527)-N(7))-methyltransferase (EC 2.1.1.170)	gene conferring resistance via absence
gyrA	DNA gyrase subunit A (EC 5.99.1.3)	antibiotic target in susceptible species; antibiotic resistant gene variant or mutant; fluoroquinolone resistance gene
gyrB	DNA gyrase subunit B (EC 5.99.1.3)	antibiotic target in susceptible species
ileS	Isoleucyl-tRNA synthetase (EC 6.1.1.5)	

Gene	Product	Classification
mexA	Multidrug efflux system, membrane fusion component => MexA of MexAB-OprM	Efflux pump conferring antibiotic resistance 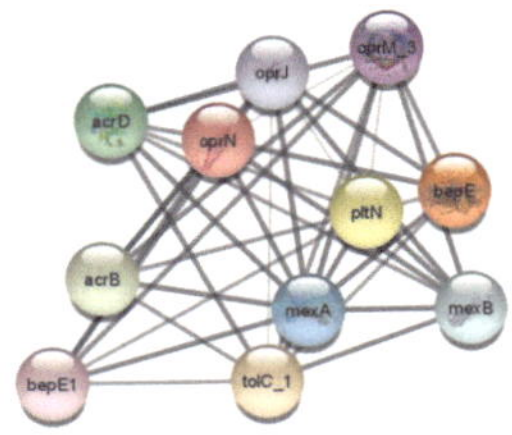
mexB	Multidrug efflux system, inner membrane proton/drug antiporter (RND type) => MexB of MexAB-OprM	
mexC	Multidrug efflux system, membrane fusion component => MexC of MexCD-OprJ system	
mexD	Multidrug efflux system, inner membrane proton/drug antiporter (RND type) => MexD of MexCD-OprJ system	
mexE	Multidrug efflux system, membrane fusion component => MexE of MexEF-OprN system	
mexF	Multidrug efflux system, inner membrane proton/drug antiporter (RND type) => MexF of MexEF-OprN system	
mexH	Multidrug efflux system, membrane fusion component => MexH of MexHI-OpmD system	
mexI	Multidrug efflux system, inner membrane proton/drug antiporter (RND type) => MexI of MexHI-OpmD system	
opmD	Multidrug efflux system, outer membrane factor lipoprotein => OpmD of MexHI-OpmD system	
oprJ	Multidrug efflux system, outer membrane factor lipoprotein => OprJ of MexCD-OprJ system	
oprM	Multidrug efflux system, outer membrane factor lipoprotein => OprM of MexAB-OprM	
oprN	Multidrug efflux system, outer membrane factor lipoprotein => OprN of MexEF-OprN system	
mexR	Multidrug resistance operon repressor MexR, MarR family	antibiotic resistant gene variant or mutant; efflux pump conferring antibiotic resistance; gene modulating antibiotic efflux
murA	UDP-N-acetylglucosamine 1-carboxyvinyltransferase (EC 2.5.1.7)	antibiotic target in susceptible species
nalC	Transcriptional regulator, AcrR family	Efflux pump conferring antibiotic resistance; gene modulating antibiotic efflux
nalD	Transcriptional regulator, AcrR family	
nfxB	Transcriptional regulator NfxB	
opdB	Outer membrane low permeability porin, OprD family => OccD7/OpdB proline	protein modulating permeability to antibiotic 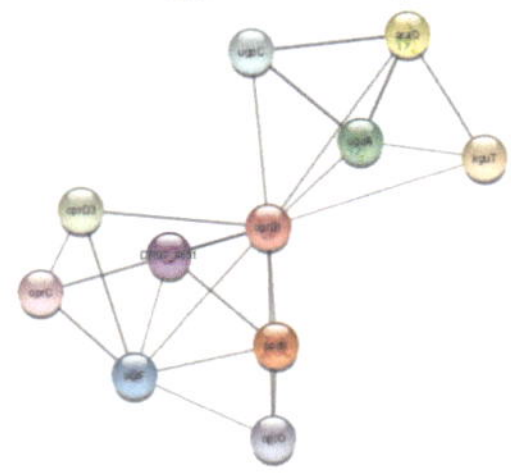
oprB	Carbohydrate-selective porin OprB	
oprD	Outer membrane low permeability porin, OprD family => OccD1/OprD basic amino acids, carbapenem uptake	
oprE	Outer membrane low permeability porin, OprD family => OccK8/OprE anaerobically-induced	
oprF	Nonspecific porin and structural outer membrane protein OprF	

Gene	Product	Classification
parC	DNA topoisomerase IV subunit A (EC 5.99.1.3)	antibiotic resistant gene variant or mutant; fluoroquinolone resistance gene; gene involved in self-resistance to antibiotic
parE	DNA topoisomerase IV subunit B (EC 5.99.1.3)	antibiotic resistant gene variant or mutant; fluoroquinolone resistance gene
pgsA	CDP-diacylglycerol–glycerol-3-phosphate 3-phosphatidyltransferase (EC 2.7.8.5)	protein altering cell wall charge conferring antibiotic resistance
phoP	Transcriptional regulatory protein PhoP	efflux pump conferring antibiotic resistance; gene altering cell wall charge conferring antibiotic resistance; gene modulating antibiotic efflux; polymyxin resistance gene
phoQ	Sensor histidine kinase PhoQ (EC 2.7.13.3)	efflux pump conferring antibiotic resistance; gene altering cell wall charge conferring antibiotic resistance; gene modulating antibiotic efflux; polymyxin resistance gene
pmrA	Two-component system response regulator QseB	gene altering cell wall charge conferring antibiotic resistance; polymyxin resistance gene
pmrB	Sensory histidine kinase QseC	
poxB	Class D beta-lactamase (EC 3.5.2.6) => OXA-50 family, oxacillin-hydrolyzing	antibiotic inactivation enzyme; beta-lactam resistance gene
rho	Transcription termination factor Rho	antibiotic target in susceptible species 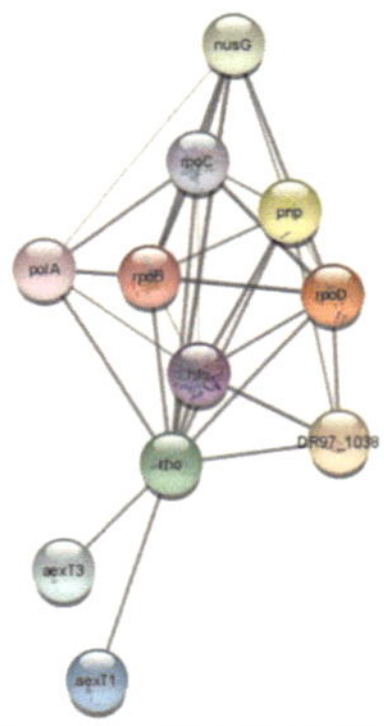
rplF	LSU ribosomal protein L6p (L9e)	
rpoB	DNA-directed RNA polymerase beta subunit (EC 2.7.7.6)	
rpoC	DNA-directed RNA polymerase beta' subunit (EC 2.7.7.6)	
rpsJ	SSU ribosomal protein S10p (S20e)	
rpsL	SSU ribosomal protein S12p (S23e)	
tufA	Translation elongation factor Tu	
tufB	Translation elongation factor Tu	
str	Aminoglycoside 6-phosphotransferase, putative	antibiotic inactivation enzyme
ParR	two-component response regulator, ParR	gene altering cell wall charge conferring antibiotic resistance
ParS	Two-component system	

Table 1.
P. aeruginosa *genes confer resistance to specific antibiotics, confer resistance to multiple antibiotics or confer multi-drug resistance (MDR); with the description of the product, and classification for each gene. Some networks in the third column were downloaded from STRING (https://string-db.org/) and analyzed with Cytoscape (https://cytoscape.org/).*

3.1 Inactivation of antibiotics by modifying enzymes

The first important mechanism that *P. aeruginosa* displays to elude the inhibitory effect of antibiotics is drug inactivation through enzymes such as aminoglycoside-modifying enzymes and β-lactamases.

3.2 The β-lactamase enzymes

The β-lactamase enzymes hydrolyze the amide bond of the four-membered β-lactam ring. The β-lactamases are classified into four classes: one zinc-dependent or metallo-β-lactamases (class B) and three active-site serine β-lactamases (classes A, C, and D). The β-lactamases of class C are widely distributed in the *P. aeruginosa* chromosome. The overexpression provides resistance to cefoxitin, cefazolin, most penicillins, cephalothin, and β-lactam inhibitor-β-lactam combinations. The AmpC gene encodes the hydrolytic enzyme β-lactamase of class C. The clinical significance of class C enzymes is enhanced by the presence of some family members, such as FOX, DHA, and CMY enzymes, on mobile genetic elements in Enterobacteriaceae and *P. aeruginosa* [20, 21].

3.3 Aminoglycoside-modifying enzymes

A common mechanism of resistance to aminoglycoside resistance in *P. aeruginosa* is the antibiotic inactivation by chromosome- or plasmid-encoded modifier enzyme. This mechanism involves three enzymes: the aminoglycoside acetyltransferase (AAC), the aminoglycoside nucleotidyl transferase (ANT), and the aminoglycoside phosphoryl transferase (APH) [22]. The AAC (6′) family of enzymes provides resistance to kanamycin, gentamicin, amikacin, netilmicin, and tobramycin.

The APH family confers inactivation by phosphorylation of aminoglycosides such as streptomycin, neomycin, and kanamycin. The *aphA*-encodes an APH (3′)-II-type enzyme, and its *aph* (3′)-IIb chromosomal aminoglycoside phosphotransferase gene confers resistance to several aminoglycosides, such as neomycin, kanamycin, amikacin, gentamicin, and tobramycin. The *aph* (3')-IIb gene forms an operon with the AraC-type positive transcriptional regulator (hpaA), and in response to the presence of 4-HPA acid, HpaA promotes the expression of *aph* (3')-IIb and that of the hpa regulon promoting the production of metabolic enzymes for the utilization of 4-HPA [22]. The ANTs family provides the adenylation of aminoglycosides such as gentamicin and streptomycin in resistant strains of *P. aeruginosa* [23, 24].

3.4 Modification of antibiotics target

In *P. aeruginosa*, the target alteration is commonly developed through mutations over the *par*C gene encoding *gyrA* gene that encoding the DNA gyrase subunit A and the DNA topoisomerase IV subunit A, both genes conferring resistance to fluoroquinolone and are homologous. In addition, 16S rRNA methyltransferases confer high levels of aminoglycoside resistance, such as rmtA, gidB, and rsmG, please refer to the methyltransferases and interactome chapter for more information about methyltransferases [25].

3.5 Global response efflux pumps conferring multidrug resistance

Multidrug efflux is an important mechanism responsible for multidrug resistance, which limits the therapeutic potential of antibiotics, failing to treat the infection caused by *P. aeruginosa***.** Multidrug efflux pumps decrease the intracellular concentration of antibiotics through a series of carrier proteins located in the bacterial cell membrane and periplasm. These proteins act by pumping various foreign substrates outside the bacterial cell, including antimicrobials, toxins, and heavy metals. The efflux pumps also export virulence determinants, including adhesins (biofilm formation) or other proteins, which are important for colonization in host cells.

The pumps can be classified according to the substrate transporting one substrate or a variety of antibiotics and other structurally different substances. In *P. aeruginosa*, there are five main families of efflux transporters: Major facilitator family (MF), ATP-binding cassette (ABC), family of cell division and nodulation of resistance (RND), family of extrusion of toxic compounds and multidrug (MATE), and small family of multidrug resistance (SMR) [26].

The RND-type multidrug efflux pumps are usually present in Gram-negative bacteria. They are integrated into a tripartite complex that includes an outer membrane channel protein, an inner membrane carrier protein, and a membrane fusion protein that joins the inner membrane protein with the outer membrane protein. In *P. aeruginosa*, several Mex efflux systems contributing to antibiotic resistance have been described, such as MexXY-OprM, MexEF-OprN, MexCD-OprJ, and MexAB-OprM.

The MexAB-OprM efflux pump belonging to the RND family is the principal contributor to multiresistance in *P. aeruginosa,* yielding resistance to a diversity of antimicrobial agents: fluoroquinolones, β-lactams, chloramphenicol, cephalosporins, macrolides, carbapenems, tetracycline, and chloramphenicol. The MexAB-OprM efflux pump is constituted by three components: a membrane fusion protein (MexA), a cytoplasmic membrane transporter (MexB), and an External membrane pore (OprM). The mexAB-oprM operon is regulated by nalC, nalD, and mexR repressor genes. Mutations in these repressor genes can lead to overexpressing of the MexAB-OprM pump causing antibiotic resistance [27–30]. The MexXY-OprM (also referred such as AmrAB) regulated by the *par*R-*par*S genes confers resistance to a broad range of antibiotics: tetracyclines, macrolides, glycylcyclines, fluoroquinolones, β-lactams, chloramphenicol, lincomycin, and novobiocin [24].

3.6 Antibiotic resistance by alterations in the cell wall charge

Polymixin B and colistin are the last antibiotic resources to treat multidrug-resistant Gram-negative bacteria, not only *P. aeruginosa*, but also *Klebsiella pneumoniae* and *Acinetobacter baumannii*. The two-component regulatory systems, PhoP-PhoQ and PmrA-PmrB, modulate resistance to cationic antimicrobial peptides, polymyxin B, colistin, and aminoglycosides. This is in response to low Mg^{2+} concentrations, regulated by an LPS modification operon arnBCADTEF-pmrE that encodes lipopolysaccharide-modifying enzymes. On the other hand, PhoP-PhoQ two-component regulatory system contributes to polymyxin resistance. Yang et al. have identified additional PhoP-regulated genes that contribute to the tolerance to polymyxin B [31].

The PhoQ works as a phosphatase that dephosphorylates the regulatory protein PhoP, binding directly to the promoter region of the oprH-phoP-phoQ operon and promoting its transcription. As a result, in the absence of the phoQ gene, genes regulated by PhoP are constitutively expressed. In addition to its operon, PhoP regulates different genes involved in antibiotic resistance, LPS modification, and antimicrobial peptides. In *P. aeruginosa*, the lipid A can be modified by enzymes encoded by the arnBCADTEF operon. These enzymes covalently add 4-amino-4-deoxy-L-arabinose (L-Ara4N) to lipid A, decreasing the negative charge of lipid A and reducing polymyxin and colistin binding. PhoP-PhoQ and other two-component regulatory systems, CprR-CprS, PmrA-PmrB, BqsR-BqsS, and ParR-ParS, directly control the expression of the arnBCADTEF operon [32–34].

3.7 Porins modulating reduction of permeability to antibiotics

P. aeruginosa expresses multiple types of porins, of which OprF (nonspecific porin and structural outer membrane protein OprF) is the most common and has been involved in diverse functions [35]. The OprD porin (outer membrane low permeability porin-basic amino acids, carbapenem uptake) contains the binding sites for carbapenems, and the absence of OprD in *P. aeruginosa,* increases resistance to carbapenems. The overexpression of the OprH porin as a result of Mg^{2+} starvation is associated with resistance to polymyxin B and gentamicin through modification of LPS.

Other important porins associated with antibiotics resistance encountered in *P. aeruginosa* are the OprB porins (carbohydrate-selective porin), OprE (outer membrane low permeability porin, OprD family), oprM (outer membrane factor lipoprotein → OprM of MexAB-OprM, multidrug efflux system), oprJ (outer membrane factor lipoprotein → OprJ of MexCD-OprJ system, multidrug efflux system), oprN (outer membrane factor lipoprotein → OprN of MexEF-OprN system, multidrug efflux system) [36–38].

4. Methyltransferases and resistome

Methyltransferase enzymes are fundamental to developing aminoglycoside resistance in *P. aeruginosa*. Some mechanisms were initially well described in *Escherichia coli*. This bacterium has 10 methylations in 16S rRNA and 14 methylations in 23S rRNA [39]. These mechanisms begin to be solved in *P. aeruginosa* [40] and other bacteria [41]. High-throughput omics such as genomics, transcriptomics, and proteomics, together with bioinformatics, data mining, and statistical analysis, are the perfect blend to identify resistome genes and protein networks.

The RNA methyltransferases carry out epigenetics methylation or demethylation on ribosomal RNA to restrict or allow antibiotic resistance development [42]. In *P. aeruginosa*, the Kam and Kgm families of methyltransferases confer resistance to several aminoglycosides [43]. Loss of native methylation may also confer a resistant phenotype. For example, when comparing the RsmG protein of *P. aeruginosa* with that of *Thermus aquaticus*, high structural homology is observed [43]. On the other hand, other methyltransferase genes have been poorly studied *in P. aeruginosa* as *gid*B (glucose-inhibited division gene). It is known that the *gid*B gene is involved in posttranslational modification methylation of 16S RNA. The *gid*B mutants also show a compromised overall bacterial fitness in *Salmonella* [44].

However, it is not only the development of aminoglycosides resistance by *P. aeruginosa* that has been significant, being a crucial case to be studied. Other methyltransferases have been associated with other antibiotics, such as macrolides comprise erythromycin and azithromycin [45]. The macrolide resistance in *Mycobacterium abscessus* complex (MABc) can develop through chromosomal mutations on the 23S rDNA (rrl) gene that promotes a high resistance level. This process goes through the induction of the *erm* gene, which causes inducible resistance in the presence of a macrolide [46]. For example, in *E. coli* studies, mobile resistance genes encoding Mph phosphotransferases or Erm methyltransferases expressed by their *mph* and *erm* genes, which were discovered in clinical settings, but their origins remain unknown. The *erm* gene from *P. aeruginosa* belongs to the large family UGF311 and confers

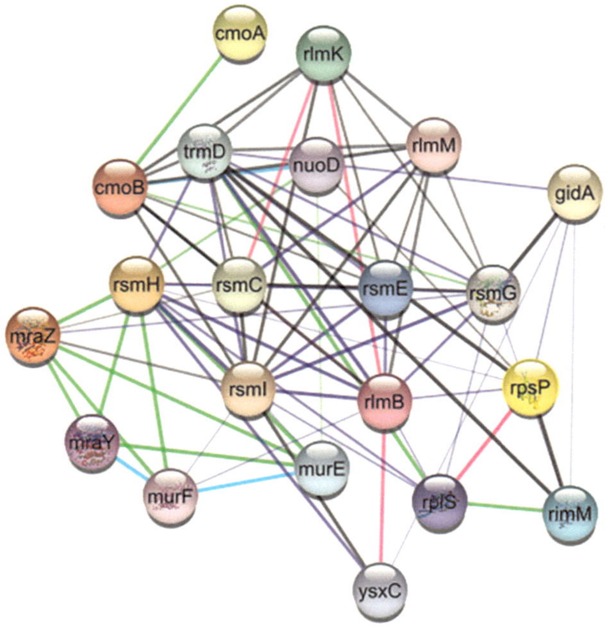

Figure 2.
Network of interactions of methyltransferases RsmH, RsmG, RsmE, RsmI, and RsmC, among others.

resistance to erythromycin and azithromycin (3,8 μg ml−1 and 3,2 μg ml−1, respectively) [45].

The RsmG, RsmH, and RsmI methyltransferases were described initially in *E. coli* with antibiotic resistance. The *rsmG* mutations confer streptomycin resistance in *E. coli*. This methyltransferase methylates the N7 of nucleotide G527 within the 530 loop of 16S rRNA [42]. This methyltransferase is associated not only with aminoglycoside resistance and the collective importance of these rRNA modifications but even involved in protein synthesis [47–49]. However, RsmG was associated with oxidative stress [50] and redox imbalance processes in *Deinococcus geothermalis* [51] and *Bacillus subtilis*, promoting low-level resistance to streptomycin, but *rsmG* mutants emerged spontaneously at a relatively high frequency, 10−6 [52]. In our studies, we did not find evidently relationships between resistance to antibiotics from *rsmG, rsmH,* and *rsmI* mutants [53].

The methyltransferase activity is diverse: antibiotic resistance, mobile genetic elements, and oxidative stress, among others. The resistome for antibiotic resistance and therefore, its interactome encloses a broad genome plasticity [54] and a high diversity of extensive proteomic interactions. For example, RsmG, RsmH, and RsmI methyltransferases interact with other proteins shown in **Figure 2**. It will be complex to establish the wide bacterial resistome; however, omics analysis is going in the right direction.

5. Interactome

The interactome is defined as the complex network of biomolecules used by a living cell or organism to interact with another living cell or organism [55]. This interaction occurs through proteins essentially, which are the signaling molecules most commonly used by cells and unicellular organisms such as bacteria. Advances in technology and bioinformatics have made it possible to map the interactome of several species, especially microorganisms, and to find out about their habitats and host adaptation mechanisms [56, 57]. There are different kinds of interactome networks: within the same species, which include protein-protein interactions, and also within different species, as in the case of interactions between bacteria. Bacteria can develop resistance and adaptation capabilities that give them advantages over their hosts due to mechanisms such as horizontal gene transfer [58, 59].

The pathogen-host interactome has been well studied from genome-wide analysis. For instance, it has been found that the versatility of many microorganisms, such as viruses, to adapt to host conditions, depends on the sequestration of endogenous host protein-protein interfaces. Like in many microorganisms, it tends to mimic or hijack host interfaces rather than generate new ones due to the limited innovative nature of microorganisms. For this reason, evolutionary studies of the interactome between species have revealed that depends on the antagonistic interactions between the pathogen and its host and the pathogen's limited capacity for innovation [60].

The study of interactome interspecies listed model organisms, such as *E. coli*, *Drosophila melanogaster*, *Saccharomyces cerevisiae*, *Caenorhabditis elegans*, and *Arabidopsis thaliana*. These species were the main tools to analyze, within a systemic approach, complex macromolecular networks, designing topological and dynamic properties that reflect biological phenomena [12]. The main networks of protein-protein interactions were built to a system-level understanding of cellular processes from genomics and proteomics data. The best example today is STRING (*Search Tool for the Retrieval of Interacting Genes/Proteins*), which contains multiple metadata from different sources, such as experimental and computational data, or statistical predictor identification and metadata of scientific collections [61]. String also contains the networks of antibiotic-resistant genes and proteins from the well-known *P. aeruginosa* PAO and ATCC 10145 strains and other strains and bacteria species.

P. aeruginosa has been used to investigate interactome too. Trouillon et al. analyzed three different strains of *P. aeruginosa* to identify Hfq-interacting RNAs, which are one the most relevant systems in the mRNA translation and degradation in bacteria. They identified growth phase-specific and strain-specific Hfq targets, including previously undescribed sRNAs, and found that the accessory Hfq interactome includes most mRNAs encoding Type III Secretion System (T3SS) components, secreted toxins, and a cluster of CRISPR guide RNAs [62].

Interactomics has a problem when the studies tend to focus on single reference strains and assume that the discoveries can be applied to the entire species, forgetting the critical intra-species genetic diversity. It will be vital to focus the interactomics studies including or enriching the number of strains to be correlated with other species. In *P. aeruginosa*, it is true that the best studies are developed with reference PAO strain; however, interesting studies have begun to change this point of view [62, 63].

6. Methyltransferases and interactome

DNA methylation is a major epigenetic factor influencing gene expression in prokaryotes and eukaryotes. DNA methylation is carried out by the enzymes known as methyltransferases, which involve the addition of a methyl group to cytosines at the

5'-position to form 5-methylcytosines [64] and also, the addition of a methyl group to adenines at the 6'-position to assemble N6-methyladenine (6mA) [65]. Dcm (DNA cytosine methyltransferase) is one of the methyltransferase enzymes discovered in prokaryotes. This, as first depicted in *E. coli*, is highly conserved in the different strains of this bacterium and reduces the expression of ribosomal protein genes during the stationary phase [66].

Methyltransferases are part of the network of biomolecules that make up the bacterial interactome. In the case of 6mA, which is part of the defense systems against phages and plasmids, it allows bacteria to distinguish between their own DNA and that of the invader, thus generating a complex network of biomolecules that allow the bacterial cell to mark (methylate) its DNA and protect it from the digestion carried out by restriction enzymes, which degrade the invading DNA that is not methylated [67].

Likewise, methyltransferases are among the molecules that play a relevant role in the interaction of bacteria with their hosts. Some studies reported that *Helicobacter pylori*, a pathogen related to the development of gastric cancer, can induce DNA hypermethylation in human gastric mucosal epithelial cells by manipulating some gene pathways associated with early steps of oncogene expression [68, 69]. Ali et al. found a link between the Rv1523 methyltransferase of *Mycobacterium tuberculosis* and its contribution to virulence, related to antibiotic resistance. In this study, they found that Rv1523 methyltransferase catalyzes the transfer of methyl group from S-adenosyl methionine (SAM) to the cell wall components of *M. tuberculosis*, modifying their components and conferring antibiotic resistance and survival under stress and acidic conditions [70].

In *P. aeruginosa*, the involvement of methyltransferases in the bacterial interaction with the host and its ability to adapt to host conditions as antibiotic pressure was largely reported. For example, the 16S rRNA methyltransferase enzyme, along with other enzymes, has been shown to be responsible for aminoglycoside resistance, acting through the irreversible transfer of methyl groups to the amino or glycoside groups of these antibiotics [71].

Bacterial epigenetic marks are currently generating a lot of interest from researchers as they are part of the bacterial interactome responsible for antibiotic resistance, for instance. The interaction of multiresistant bacteria with their hosts can cause molecular alterations in the transcriptional programs of host cells involving epigenetic mechanisms such as non-coding RNAs, DNA methylation, and histone modifications conferring survival to the bacteria [72]. However, more studies are still needed to understand in depth the epigenetic mechanisms that are part of the bacterial interactome to find new therapeutic targets to address public health problems such as bacterial multiresistance to antibiotics.

7. Relationship resistome-pangenome

What is the size of the resistome inside the pangenome? And what is the set of resistome genes? These questions will be the future challenge in the antibiotic's resistance panorama and its relationship with other omics such as pangenomics. The term "pangenome"/pangenome was created by Tettelin et al. in 2005 [73] to describe the complete set of genes of a given species by sequencing the complete genomes of several strains of that species. The pangenome is composed of the core genome, accessory genome, and unique genome.

The core or central genome refers to the set of genes common to all the genomes of the species studied. The accessory or variable genome refers to the collection of genes that are in one or various genomes but not in all of them in each species studied. While the core genome is highly conserved, the second serves to understand the variations in the genomes of the different species studied; thus, their specific lifestyles and evolutionary trajectories. Finally, the unique genome refers to the set of genes with no homologous copy within the genomes under study [16, 74, 75].

The resistome is part of the pangenome and is the set of genes that confer resistance to a bacterial species and that are transmitted from one strain to another by different mechanisms known as part of the global set of genes of that bacterial species. Indeed, according to Gillings, the genes that confer antibiotic resistance are ancient components of the bacterial pangenome, having been recovered from 30,000-year-old permafrost and from a cave microbiome of 4 million years that has been isolated [76]. Therefore, pangenomics analysis is one of the tools available today for in-depth and broad studies of the resistome in pathogenic microorganisms [77].

However, pangenomics faces some fundamental concerns: actually, today it is impossible to solve all the questions using classical molecular biology or descriptive genomics. The core and accessory genome help to figure out the role of some genes inside species. Not all resistance genes are in the core genome, and some strains are part of HGT genes as unique genes, never incorporated in the core genome, giving plasticity to the resistome. For this reason, the study of the resistome through pangenomics analyses remains a challenge for current research that requires new developments to better realize the complex landscape of resistance genes [78, 79].

8. Conclusion

The opportunistic pathogen *P. aeruginosa* has been widely studied as one of the microorganisms prioritized for research and development of new therapies to treat nosocomial infections, given its high capacity to generate resistance to currently available antibiotics, adapting easily to environmental conditions. Although, the identification of the gene that confers antibiotic resistance to *P. aeruginosa* has been a complex task. It is known that there is a genetic battery that is part of its resistome, composed of genes such as CARB-3, CARB-4, PSE-1 (CARB-2), PSE-4 (CARB-1), OXA-18, OXA-2, OXA-21, OXA-10 (PSE-2), GyrA, GyrB, OprM, OprJ, OprN, MexB, MexD, MexF, and MexY. Other less studied genes encoding methyltransferase enzymes are also part of the *P. aeruginosa* resistome. The research of the *P. aeruginosa* resistome, as well as its interactome, has advanced thanks to the development of new sequencing technologies, within which pangenomics analysis has opened a horizon to improve the understanding of the genetic mechanisms underlying antibiotic resistance. Through this analysis, it is possible to identify common and unique genes within different strains of *P. aeruginosa*. The study of the resistome is part of the pangenomics analysis. Nevertheless, a thorough understanding of the *P. aeruginosa* resistome, its interactome, and other mechanisms of antibiotic resistance remains a challenge for researchers and a key element to consider in public health issues.

Acknowledgements

We thank the University of Antioquia for their financial support to the project CODI 2017-15753.

Conflicts of interest

The authors declare that they have no conflicts of interest regarding the publication of this paper.

Author details

Diana Carolina Castaño, Juan David Patiño-Salazar and Mauricio Corredor*
Gebiomic Group, Institute of Biology, University of Antioquia, Medellin, Colombia

*Address all correspondence to: mauricio.corredor@udea.edu.co

References

[1] World Health Organization. Guidelines for the prevention and control of carbapenem-resistant *Enterobacteriaceae*, *Acyinetobacter baumannii* and *Pseudomonas aeruginosa* in health care facilities. 2017. https://apps.who.int/iris/handle/10665/259462. Licencia: CC BY-NC-SA 3.0 IGO

[2] Appanna VD, Sarabhai S, Sharma P, Capalash N. Ellagic acid derivatives from Terminalia chebula Retz. Downregulate the expression of quorum sensing genes to attenuate Pseudomonas aeruginosa PAO1 virulence. PLoS One. 2013;**8**(1):e53441

[3] Kipnis E, Sawa T, Wiener-Kronish J. Targeting mechanisms of *Pseudomonas aeruginosa* pathogenesis. Médecine et Maladies Infectieuses. 2006;**36**(2):78-91

[4] Wu DC, Chan WW, Metelitsa AI, Fiorillo L, Lin AN. *Pseudomonas* skin infection. American Journal of Clinical Dermatology. 2011;**12**(3):157-169

[5] Zowawi HM, Harris PNA, Roberts MJ, Tambyah PA, Schembri MA, Pezzani MD, et al. The emerging threat of multidrug-resistant Gram-negative bacteria in urology. Nature Reviews Urology. 2015;**12**(10):570-584

[6] Rahme LG, Stevens EJ, Wolfort SF, Shao J, Tompkins RG, Ausubel FM. Common virulence factors for bacterial pathogenicity in plants and animals. Science. 1995;**268**(5219):1899-1902

[7] Chattopadhyay I. Antibiotic-resistant bacteria and antibiotic resistance genes profile in wastewater treatment plants by using genomic approaches. In: Wastewater Treatment Reactors. Edited By: Maulin P. Shah and Susana Rodriguez-Couto. Elsevier; 2021. pp. 567-582

[8] Chadha J, Harjai K, Chhibber S. Revisiting the virulence hallmarks of *Pseudomonas aeruginosa*: A chronicle through the perspective of quorum sensing. Environmental Microbiology. 2022;**24**(6):2630-2656

[9] Hendrix H, Zimmermann-Kogadeeva M, Zimmermann M, Sauer U, De Smet J, Muchez L, et al. Metabolic reprogramming of *Pseudomonas aeruginosa* by phage-based quorum sensing modulation. Cell Reports. 2022;**38**(7):110372

[10] Muhammad JS, Khan NA, Maciver SK, Alharbi AM, Alfahemi H, Siddiqui R. Epigenetic-mediated antimicrobial resistance: Host versus pathogen epigenetic alterations. Antibiotics. 2022;**11**(6):809

[11] Torrens G, van der Schalk TE, Cortes-Lara S, Timbermont L, del Barrio-Tofiño E, Xavier BB, et al. Susceptibility profiles and resistance genomics of *Pseudomonas aeruginosa* isolates from European ICUs participating in the ASPIRE-ICU trial. Journal of Antimicrobial Chemotherapy. 2022;**77**(7):1862-1872

[12] Cusick ME, Klitgord N, Vidal M, Hill DE. Interactome: Gateway into systems biology. Human Molecular Genetics. October 15, 2005;**14**(suppl_2):R171-R181

[13] Wright GD. The antibiotic resistome: The nexus of chemical and genetic diversity. Nature Reviews Microbiology. 2007;**5**(3):175-186

[14] Lee AJ, Doing G, Neff SL, Reiter T, Hogan DA, Greene CS. Compendium-wide analysis of *P. aeruginosa* core and accessory genes reveal more nuanced transcriptional patterns. bioRxiv. April 15, 2022. pp. 1-35

[15] Trouillon J, Han K, Attré I, Lory S. The core and accessory Hfq interactomes

across *Pseudomonas aeruginosa* lineages. Nature Communications. 2022;**13**(1): 1-16

[16] Corredor M, Muñoz-Gómez A. Pan-genomics of multidrug-resistant human pathogenic bacteria and their resistome. Pan-genomics: Applications, Challenges, and Future Prospects. 2020:203-235

[17] Hallinan JS, James K, Wipat A. Network approaches to the functional analysis of microbial proteins. Advances in Microbial Physiology. 2011;**59**:101-133

[18] Zhang M, Su S, Bhatnagar RK, Hassett DJ, Lu LJ. Prediction and analysis of the protein interactome in *Pseudomonas aeruginosa* to enable network-based drug target selection. PLoS One. 2012;**7**(7):e41202

[19] Hwang S, Kim CY, Ji SG, Go J, Kim H, Yang S, et al. Network-assisted investigation of virulence and antibiotic-resistance systems in *Pseudomonas aeruginosa*. Scientific Reports. 2016, 2016;**6**(1):1-11

[20] Tooke CL, Hinchliffe P, Bragginton EC, Colenso CK, Hirvonen V, Takebayashi Y, et al. β-Lactamases and β-Lactamase inhibitors in the 21st Century. Journal of Molecular Biology. 2019;**431**(18):3472-3500

[21] Jacoby GA. AmpC beta-lactamases. Clinical Microbiology Reviews. 2009; **22**(1):161-182

[22] Vaziri F, Peerayeh SN, Nejad QB, Farhadian A. The prevalence of aminoglycoside-modifying enzyme genes (aac (6')-I, aac (6')-II, ant (2'')-I, aph (3')-VI) in *Pseudomonas aeruginosa*. Clinics (São Paulo, Brazil). 2011;**66**(9): 1519-1522

[23] Zeng L, Jin S. aph(3')-IIb, a gene encoding an aminoglycoside-modifying enzyme, is under the positive control of surrogate regulator HpaA. Antimicrobial Agents and Chemotherapy. 2003;**47**(12): 3867-3876

[24] Poole K. Aminoglycoside resistance in *Pseudomonas aeruginosa*. Antimicrobial Agents and Chemotherapy. 2005;**49**(2): 479-487

[25] El Zowalaty ME, Al Thani AA, Webster TJ, El Zowalaty AE, Schweizer HP, Nasrallah GK, et al. *Pseudomonas aeruginosa*: Arsenal of resistance mechanisms, decades of changing resistance profiles, and future antimicrobial therapies. Future Microbiology. 2015;**10**(10): 1683-1706

[26] Webber MA, Piddock LJV. The importance of efflux pumps in bacterial antibiotic resistance. Journal of Antimicrobial Chemotherapy. 2003; **51**(1):9-11

[27] Aguilar-Rodea P, Zúñiga G, Cerritos R, Rodríguez-Espino BA, Gomez-Ramirez U, Nolasco-Romero CG, et al. Nucleotide substitutions in the mexR, nalC and nalD regulator genes of the MexAB-OprM efflux pump are maintained in *Pseudomonas aeruginosa* genetic lineages. PLoS One. 2022;**17**(5):e0266742

[28] Huang L, Wu C, Gao H, Xu C, Dai M, Huang L, et al. Bacterial multidrug efflux pumps at the frontline of antimicrobial resistance: An overview. Antibiotics. 2022;**11**(4):520

[29] Neuberger A, Du D, Luisi BF. Structure and mechanism of bacterial tripartite efflux pumps. Research in Microbiology. 2018;**169**(7-8):401-413

[30] Rashid Mahmood A, Mansour HN. Study of antibiotic resistant genes in *Pseudomonas aeruginosa* isolated from

burns and wounds. Archives of Razi Institute. 2022;**77**(1):403-411

[31] Yang B, Liu C, Pan X, Fu W, Fan Z, Jin Y, et al. Identification of novel phoP-phoQ regulated genes that contribute to polymyxin B tolerance in *Pseudomonas aeruginosa*. Microorganisms. 2021; **9**(2):344

[32] McPhee JB, Lewenza S, Hancock RE. Cationic antimicrobial peptides activate a two-component regulatory system, PmrA-PmrB, that regulates resistance to polymyxin B and cationic antimicrobial peptides in *Pseudomonas aeruginosa*. Molecular Microbiology. 2003;**50**(1): 205-217

[33] Ly NS, Yang J, Bulitta JB, Tsuji BT. Impact of two-component regulatory systems PhoP-PhoQ and PmrA-PmrB on colistin pharmacodynamics in *Pseudomonas aeruginosa*. Antimicrobial Agents and Chemotherapy. 2012;**56**(6): 3453-3456

[34] Macfarlane E, Kwasnicka A, Hancock R. Role of *Pseudomonas aeruginosa* PhoP-phoQ in resistance to antimicrobial cationic peptides and aminoglycosides. Microbiology. 2000;**146**(Pt 10): 2543-2554

[35] Chevalier S, Bouffartigues E, Bodilis J, Maillot O, Lesouhaitier O, Feuilloley MG, et al. Structure, function and regulation of *Pseudomonas aeruginosa* porins. FEMS Microbiology Reviews. 2017;**41**(5):698-722

[36] Langendonk RF, Neill DR, Fothergill JL. The building blocks of antimicrobial resistance in *Pseudomonas aeruginosa*: Implications for current resistance-breaking therapies. Frontiers in Cellular and Infection Microbiology. 2021;**11**:665759

[37] Pang Z, Raudonis R, Glick BR, Lin TJ, Cheng Z. Antibiotic resistance in *Pseudomonas aeruginosa*: Mechanisms and alternative therapeutic strategies. Biotechnology Advances. 2019;**37**(1): 177-192

[38] Hamed DM. Antibiotics profile and b-lactamases in *Pseudomonas aeruginosa*. Records of Pharmaceutical and Biomedical Sciences. 2021;**5**:74-79

[39] Andersen NM, Douthwaite S. YebU is a m5C methyltransferase specific for 16 S rRNA nucleotide 1407. Journal of Molecular Biology. 2006;**359**: 777-786

[40] Doberenz S, Eckweiler D, Reichert O, Jensen V, Bunk B, Spröer C, et al. Identification of a *Pseudomonas aeruginosa* PAO1 DNA methyltransferase, its targets, and physiological roles. mBio. 2017;**8**(1): e02312-e02316

[41] Moradi J, Fathollahi M, Halimi S, Alvandi A, Abiri R, Vaziri S, et al. Characterization of resistome in *Lactobacillus* genomic sequences from the human gut. Journal of Global Antimicrobial Resistance. 2022;(22): S2213, 00119-9-S7165

[42] Valderrama-Carmona P, Cuartas JH, Carolina-Castaño D, Corredor M. The role of *Pseudomonas aeruginosa* RNA methyltransferases in antibiotic resistance. In *Pseudomonas aeruginosa*, An Armory Within. Editor Dinesh Sriramulu. Intechopen. 2019. pp. 1-17

[43] Savic M, Lovrić J, Tomic TI, Vasiljevic B, Conn GL. Determination of the target nucleosides for members of two families of 16S rRNA methyltransferases that confer resistance to partially overlapping groups of

aminoglycoside antibiotics. Nucleic Acids Research. 2009;**37**(16):5420-5431

[44] Rybenkov VV. Maintenance of chromosome structure in *Pseudomonas aeruginosa*. FEMS Microbiology Letters. 2014;**356**(2):154-165

[45] Lund D, Kieffer N, Parras-Moltó M, Ebmeyer S, Berglund F, Johnning A, et al. Large-scale characterization of the macrolide resistome reveals high diversity and several new pathogen-associated genes. Microbial Genomics. 2022;**8**(1):000770

[46] van den Bogaart L, Manuel O. Antibiotic therapy for difficult-to-treat infections in lung transplant recipients: A practical approach. Antibiotics. 2022; **11**(5):612

[47] Green R, Noller HF. Reconstitution of functional 50S ribosomes from in vitro transcripts of *Bacillus stearothermophilus* 23S rRNA. Biochemistry. 1999;**38**:1772-1779

[48] Khaitovich P, Tenson T, Kloss P, Mankin AS. Reconstitution of functionally active *Thermus aquaticus* large ribosomal subunits with in vitro-transcribed rRNA. Biochemistry. 1999; **38**:1780-1788

[49] Krzyzosiak W, Denman R, Nurse K, Hellmann W, Boublik M, Gehrke CW, et al. In vitro synthesis of 16S ribosomal RNA containing single base changes and assembly into functional 30S ribosome. Biochemistry. 1987;**26**: 2353-2364

[50] Babosan A, Fruchard L, Krin E, Carvalho A, Mazel D, Baharoglu Z. Non-essential tRNA and rRNA modifications impact the bacterial response to sub-MIC antibiotic stress. BioRxiv. February 06, 2022. pp. 1-32

[51] Ye Q, Lee C, Shin E, Lee SJ. Influence of redox imbalances on the transposition of insertion sequences in *Deinococcus geothermalis*. Antioxidants. 2021;**10**(10):1623

[52] Nishimura K, Johansen SK, Inaoka T, Hosaka T, Tokuyama S, Tahara Y, et al. Identification of the RsmG methyltransferase target as 16S rRNA nucleotide G527 and characterization of *Bacillus subtilis* rsmG mutants. Journal of Bacteriology. 2007;**189**(16):6068-6073

[53] Mosquera-Rendón J, Rada-Bravo AM, Cárdenas-Brito S, Corredor M, Restrepo-Pineda E, Benítez-Páez A. Pangenome-wide and molecular evolution analyses of the *Pseudomonas aeruginosa* species. BMC Genomics. 2016;**17**(1):1-14

[54] Abril D, Marquez-Ortiz RA, Castro-Cardozo B, Moncayo-Ortiz JI, Olarte-Escobar NM, Corredor-Rozo ZL, et al. Genome plasticity favours double chromosomal Tn4401b-blaKPC-2 transposon insertion in the *Pseudomonas aeruginosa* ST235 clone. BMC Microbiology. 2019;**19**(1):1-12

[55] Ghadie MA, Coulombe-Huntington J, Xia Y. Interactome evolution: Insights from genome-wide analyses of protein-protein interactions. Current Opinion in Structural Biology. 2018;**50**:42-48

[56] Cafarelli TM, Desbuleux A, Wang Y, Choi SG, De Ridder D, Vidal M. Mapping, modeling, and characterization of protein-protein interactions on a proteomic scale. Current Opinion in Structural Biology. 2017;**44**:201-210

[57] Wodak SJ, Vlasblom J, Turinsky AL, Pu S. Protein-protein interaction networks: The puzzling riches. Current Opinion in Structural Biology. 2013; **23**(6):941-953

[58] Kadibalban AS, Bogumil D, Landan G, Dagan T. DnaK-Dependent accelerated evolutionary rate in prokaryotes. Genome Biology and Evolution. 2016;**8**(5): 1590-1599

[59] Franzosa EA, Garamszegi S, Xia Y. Toward a three-dimensional view of protein networks between species. Frontiers in Microbiology. 2012; **3**:428

[60] Daugherty MD, Malik HS. Rules of engagement: Molecular insights from host-virus arms races. Annual Review of Genetics. 2012;**46**:677-700

[61] Szklarczyk D, Franceschini A, Kuhn M, Simonovic M, Roth A, Minguez P, et al. The STRING database in 2011: Functional interaction networks of proteins, globally integrated and scored. Nucleic Acids Research. 2011;**39** (Database issue):D561-D568

[62] Trouillon J, Imbert L, Villard AM, et al. Determination of the two-component systems regulatory network reveals core and accessory regulations across Pseudomonas aeruginosa lineages. Nucleic Acids Research. 2021;**49**(20): 11476-11490

[63] Gebhardt MJ, Kambara TK, Ramsey KM, Dove SL. Widespread targeting of nascent transcripts by RsmA in *Pseudomonas aeruginosa*. Proceedings of the National Academy of Sciences of the United States of America. 2020; **117**(19):10520-10529

[64] van der Wijst MG, Venkiteswaran M, Chen H, et al. Local chromatin microenvironment determines DNMT activity: From DNA methyltransferase to DNA demethylase or DNA dehydroxymethylase. Epigenetics. 2015;**10**:671-676

[65] Oerum S, Meynier V, Catala M, Tisné C. A comprehensive review of m6A/m6Am RNA methyltransferase structures. Nucleic Acids Research. 2021; **49**(13):7239-7255

[66] Militello KT, Simon RD, Qureshi M, et al. Conservation of Dcm-mediated cytosine DNA methylation in *Escherichia coli*. FEMS Microbiology Letters. 2012;**328**(1):78-85

[67] Low DA, Weyand NJ, Mahan MJ. Roles of DNA adenine methylation in regulating bacterial gene expression and virulence. Infection and Immunity. 2001;**69**:7197-7204

[68] Yao Y, Tao H, Park DI, et al. Demonstration and characterization of mutations induced by *Helicobacter pylori* organisms in gastric epithelial cells. Helicobacter. 2006;**11**:272-286

[69] Ando T, Yoshida T, Enomoto S, et al. DNA methylation of microRNA genes in gastric mucosae of gastric cancer patients: Its possible involvement in the formation of epigenetic field defect. International Journal of Cancer. 2009; **124**:2367-2374

[70] Ali S, Ehtram A, Arora N, Manjunath P, Roy D, Ehtesham NZ, et al. The *M. tuberculosis* Rv1523 methyltransferase promotes drug resistance through methylation-mediated cell wall remodeling and modulates macrophages immune responses. Frontiers in Cellular and Infection Microbiology. 2021;**11**: 622487

[71] Behzadi P, Baráth Z, Gajdács M. It's not easy being green: A narrative review on the microbiology, virulence and therapeutic prospects of multidrug-resistant *Pseudomonas aeruginosa*. Antibiotics (Basel, Switzerland). 2021; **10**(1):42

[72] Crimi E, Benincasa G, Cirri S, et al. Clinical epigenetics and multidrug-resistant bacterial infections: Host remodelling in critical illness. Epigenetics. 2020;**15**(10): 1021-1034

[73] Tettelin H, Masignani V, Cieslewicz MJ, Donati C, Medini D, Ward NL, et al. Genome analysis of multiple pathogenic isolates of Streptococcus agalactiae: Implications for the microbial "pangenome". Proceedings of the National Academy of Sciences of the United States of America. 2005;**102**:13950-13955

[74] Her HL, Lin PT, Wu YW. PangenomeNet: A pan-genome-based network reveals functional modules on antimicrobial resistome for *Escherichia coli* strains. BMC Bioinformatics. 2021; **22**(1):548

[75] Rouli L, Merhej V, Fournier PE, Raoult D. The bacterial pangenome as a new tool for analysing pathogenic bacteria. New Microbes New Infections. 2015;7:72-85

[76] Gillings MR. Evolutionary consequences of antibiotic use for the resistome, mobilome and microbial pangenome. Frontiers in Microbiology. 2013;**4**(4):1-10

[77] Aggarwal SK, Singh A, Choudhary M, Kumar A, Rakshit S, Kumar P, et al. Pangenomics in microbial and crop research: Progress, Applications, and perspectives. Genes (Basel). 2022;**13**(4):598

[78] Horesh G, Taylor-Brown A, McGimpsey S, Lassalle F, Corander J, Heinz E, et al. Different evolutionary trends form the twilight zone of the bacterial pan-genome. Microbial Genomics. 2021;7(9):000670

[79] Urhan A, Abeel T. A comparative study of pan-genome methods for microbial organisms: *Acinetobacter baumannii* pan-genome reveals structural variation in antimicrobial resistance-carrying plasmids. Microbial Genomics. 2021;**7**(11):000690

Chapter 3

Pseudomonas aeruginosa: The Alarming Pathogen of Hospital Acquired Infection

Anika Farzin, Md. Mizanur Rahman and Ferdows Ara Mollika

Abstract

Pseudomonas aeruginosa is a Gram-negative bacillus that is ubiquitous. It is found in abundant amount in the environment, especially in moist places. Immunocompetent persons get less infected by *P. aeruginosa*, while immunocompromised patients get more infected by them. The burn patients, patients with cystic fibrosis, and patients who are dependent on any device like ventilator, intravenous catheter, or indwelling bladder catheter are more prone to acquire the infection. It is the main cause of ventilator-associated pneumonia. Patients with neutropenia are more susceptible to *P. aeruginosa* infections. *P. aeruginosa* has got several virulence factors through which they can cause disease. They have got attachment factors like Pilli and enzymes. They are one of the main pathogens of healthcare-associated infection (HAI). As *P. aeruginosa* is multidrug-resistant so this has got an extra contribution to the fact that they can cause HAI more. Common antibiotics like penicillin, carbapenems, cephalosporins, and all other Beta-lactam drugs along with aztreonams and fluroquinolones are resistant to *P. aeruginosa*. The proper maintenance of hand hygiene and continuous monitoring of hospital devices can lessen the burden of HAI associated with *P. aeruginosa* infection.

Keywords: *Pseudomonas aeruginosa*, host defense, urinary catheter, antimicrobial resistance, healthcare-associated infection, nosocomial infection

1. Introduction

Nosocomial infections (NI) are a worldwide problem in hospital care. This is an important obstacle that worsens the prognosis of the underlying disease, increases mortality, prolongs hospitalization, worsens the quality of life of patients, and increases the expenses of treatment, so NI needs to be paid special attention [1]. CDC the leading expert of infectious disease control in the United States has set procedures and guidelines that are most widely used worldwide. A nationwide NI surveillance system in the United States has been organized since the 1970s [2].

The United Kingdom also has laboratory service of the British Public Health Service (Public Health Laboratory Service) to monitor the control of healthcare-associated infection (HAI). The main sponsor, which organizes assemblies dedicated to HAI, is the Hospital Infection Society (HIS), that publishes a globally significant and standard journal—the *Journal of Hospital Infection* [3]. HAIs need to be diagnosed and treated in time, but the most important thing is to maintain proper precaution in various hospital wards, especially in intensive care units (ICU) [4]. Infections developed after hospitalization can lead to remarkable amount of morbidity and mortality, but protective measures can significantly lessen the rate [5]. It is necessary to prevent healthcare individuals to reduce the chances of getting infected and spreading it to other patients and hospital staffs. Thus, the absence of hospital staff from going to work can be reduced, which can improve the quality of the skills of the health providers of ICU. Nosocomial infections are also allied with the burden of extra expenses [6].

Pseudomonas aeruginosa has been detected as a pathogen of hospital patients only in the modern era of intensive treatment and antibiotic resistance. *P. aeruginosa* causes 10–11% of all NI. It is a Gram-negative, motile, aerobic rod, widely distributed in nature and commonly present in moist environments with simple nutrients in hospitals. *P. aeruginosa* has comparative resistance to antibiotics and disinfectants which allows it to become established often in very large numbers in fluids, and wet places in hospitals and also to colonize the mucous membranes and skin of patients. In 2017, multidrug-resistant *Pseudomonas aeruginosa* caused an estimated 32,600 infections among hospitalized patients and 2700 estimated deaths in the United States [7].

2. Compromised host

A compromised host is that individual whose resistance to infection is reduced by disease, therapy, or burns. The host can be compromised by two principal conditions: injured skin or mucous membranes, and a compromised immune system.

The skin, skin appendages, and mucous membranes provide challenging physical barriers in counter to most pathogens. Burn injuries, surgical wounds, trauma (such as accidental wounds), injections, invasive diagnostic procedures, ventilators, intravenous therapy, and urinary catheters can break the first line of defense; thus, making a person more susceptible to disease in a hospital.

Burn patients are especially susceptible to hospital-acquired infections because their skin is no longer an effective physical barrier to microorganisms. Patients requiring invasive procedures usually suffer from serious diseases, which further increases susceptibility to infections. Invasive devices provide a pathway for microorganisms in the environment to enter the body and also help transfer them from one part of the body to another. Pathogens can also proliferate on the devices themselves [4].

It causes disease in humans with abnormal host defenses, especially in patients with neutropenia. In healthy individuals, white blood cells known as T lymphocytes provide resistance to disease by killing pathogens directly, mobilizing phagocytes and other lymphocytes, and B lymphocytes develop into antibody-producing cells to protect against infection. Antibodies neutralize the toxins, inhibit adherence of a pathogen to host cells, and thus kill the pathogens. Drugs, ionizing radiation, steroids, burn injuries, diseases like diabetes, leukemia, kidney disease, mental conditions like stress, and malnutrition can all unpleasantly affect the actions of T lymphocytes and B lymphocytes to compromise the host. Again, HIV affects certain cells which are important for our immune system [8].

It is usually pathogenic when introduced into areas devoid of normal defenses when intravenous or urinary catheters are used. They produce systemic diseases with the help of pilus, enzymes, glycocalyx, and toxins.

Pseudomonas aeruginosa has large intrinsic resistance to multiple antibiotics. This characteristic and its quick ability to acquire new antimicrobial resistance make this pathogen a growing problem in infectious disease pathology, especially when nosocomial in origin. Traditionally *P. aeruginosa* is not treated with single-drug therapy because the success rate is very low and the bacteria can rapidly develop resistance when a single drug is used. Multidrug resistance has become a major issue in the management of hospital-acquired urinary tract infections [7].

3. Antigenic structure and toxins

Pseudomonas aeruginosa specifically produces a variety of virulence factors including adhesins, enzymes, and toxins. Pili extends from the cell surface and help in attachment with host epithelial cells. Lipopolysaccharide, an element of the cell wall structure is responsible for many of the endotoxic properties of the organism. Most of the *Pseudomonas aeruginosa* isolates from clinical infections produce extracellular enzymes, including elastases, proteases, and two hemolysins- a heat-labile phospholipase C and a heat-stable glycolipid. A pigment known as pyocyanin produced by *P. aeruginosa* is responsible for the production of hydrogen peroxide and superoxide and stimulates IL-8. The increased release of IL-8 serves as an attractant for neutrophils. The other pigment pyoverdine is a siderophore; it binds to iron.

Pseudomonas aeruginosa produces exotoxin A which causes tissue necrosis and is lethal for animals when injected in purified form. The toxin blocks protein synthesis. *P. aeruginosa* produces four type III secreted toxins; exoenzyme S, exoenzyme T, exoenzyme U, and exoenzyme Y. Among these, exoenzymes S and exoenzyme T are enzymes with GTPase and ADP-ribosyl transferase activity whereas Exoenzyme U is a phospholipase and Exoenzyme Y is an adenylyl cyclase [9]. These toxins cause cell death or inhibit the host's response to infection.

4. Clinical diseases

4.1 Pulmonary infection

Pseudomonas aeruginosa causes infection of the lower respiratory tract that can differ in severity from asymptomatic colonization or benign inflammation of the bronchioles to severe necrotizing bronchopneumonia. Colonization can be seen in patients with cystic fibrosis, other chronic lung diseases, and patients who have neutropenia. Infections in patients with cystic fibrosis have been connected with exacerbation of the underlying disease and invasive pulmonary disease. Mucoid strains are associated with these strains and are difficult to remove because chronic infections with these bacteria are associated with a progressive increase in antibiotic resistance.

Nosocomial pneumonia is the second most common nosocomial infection in pediatric ICUs after intravenous catheter infections of the bloodstream. It can occur in any patient but is most common in infants, young children, and patients over 65 years of age which means they are immunocompromised. Patients in pediatric ICUs who are most at risk for pneumonia are patients who have been intubated and are under mechanical ventilation. The risk intensifies as vocal cords remain open and there is an increased risk of aspiration of gastrointestinal contents. The chances of developing nosocomial pneumonia are 6–20 times higher in patients who undergo mechanical ventilation rather than those who do not need that. Ventilator-associated pneumonia (VAP) is defined as a lung infection resulting in pneumonia that occurs in people who are on mechanical ventilation for at least 48 hours.

Factors that can initiate the development of VAP in children are immunodeficiency, immunosuppression, and neuromuscular blockade. The other factors are myasthenia gravis, burn injuries, administration of steroids, and hyperalimentation. Children who are at higher risk of VAP are those with antibiotics, with a prolonged stay in the ICU, with intravenous channels who have greater risk of hematogenous spread, H2-receptor blocker intake, reinsertion of endotracheal tube, and transport of patient outside during intubation as they may get contaminated by the environment. The presence of a nasogastric tube provides a direct pathway from the upper gastrointestinal tract to the oropharynx.

The risk of nosocomial pneumonia will be increased by the use of nebulizers and manipulation of the ventilator circuit. VAP in children accounts for 10–26% of nosocomial infections. Nosocomial pneumonia accounts for a 20–70% mortality rate which is highest in all pediatric nosocomial infections, although the period of endotracheal intubation increases the risk of nosocomial pneumonia, which is highest during the first 2 weeks of intubation.

Within 5 days almost all the children who have received endotracheal tube gets colonized by nosocomial microorganisms. The most commonly identified bacteria are gram-negative bacilli, especially *P. aeruginosa*, in which mortality is higher. One of the leading causes of ventilator-associated pneumonia (VAP) in the United States and Europe [10] is *P. aeruginosa*, which poses unique challenges for its clinical management and increase in incidence.

4.2 Urinary tract infections

Nosocomial infection accounts for 20–49% in all the cases of urinary tract infection, out of which 10% is contributed by *Pseudomonas aeruginosa* which usually occurs after catheterization, instrumentation of the urinary system or surgery, resulting in discomfort, pyelonephritis, morbidity, and in some cases, it may lead to death. Insertion of catheter can disrupt the mucosal lining which may lead to easy entry of pathogens into the urinary tract causing bacterial colonization [11].

In UTI *Pseudomonas aeruginosa* is a key uropathogen with high prevalence in reported cases worldwide [12].

4.3 Skin and soft tissue infection

P. aeruginosa can cause a variety of primary skin infections and the most recognized among them is the burn wound. Colonization of a burn wound followed by localized vascular damage, tissue necrosis, and ultimately bacteremia. The moist surface of the burns and the inability of neutrophils to enter the wound predispose the patients to such infections.

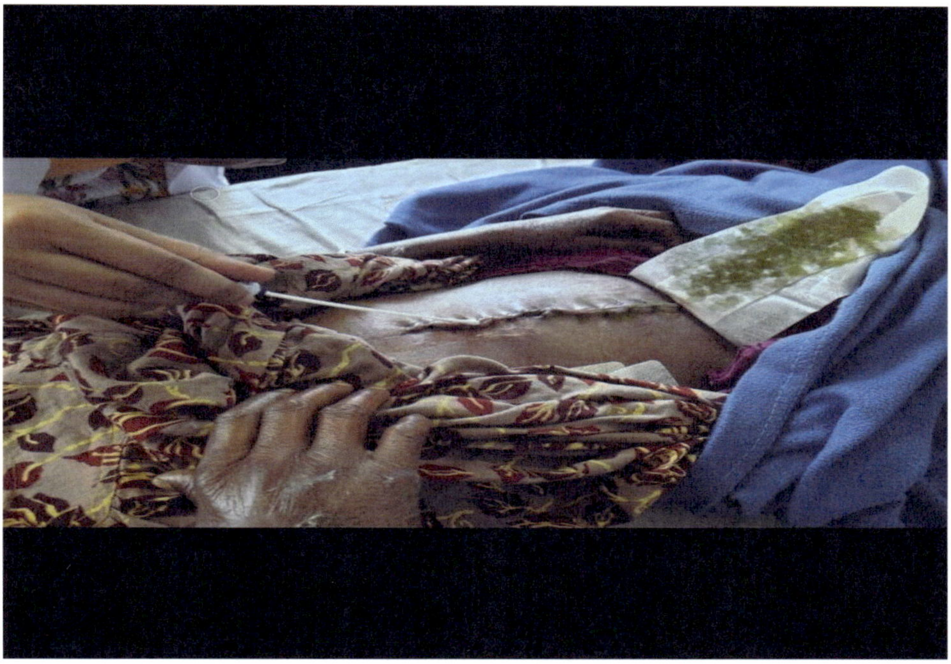

Picture showing the typical greenish pus of *Pseudomonas aeruginosa* in a patient with surgical wound infection.

4.4 Bacteremia

Bacteremia may also occur due to *P. aeruginosa* which is a cause of high mortality rate. Bacteremia occurs most often in patients with neutropenia, diabetes mellitus, extensive burns, and malignancies. In most cases, they originate from the lower respiratory tract, urinary tract, and skin and soft tissue infections [13].

4.5 Antimicrobial resistance of Pseudomonas aeruginosa

Antimicrobial resistance happens when germs like bacteria and fungi gain the ability to resist the action of an antimicrobial agent that was designed to kill them, to which it was previously susceptible [14]. HAI caused by antibiotic-resistant *P. aeruginosa* has appeared as an alarming concern in health care settings as day-by-day MDR strains (i.e., resistance to at least three antibiotics) are increasing. *P. aeruginosa* shows resistance to various antimicrobial agents because of its outer membrane with low permeability. Although some other mechanisms are also responsible for their intrinsic resistance including, the efflux system which expels antibiotics out of the bacterial cell, and the production of the antibiotic-inactivating enzyme.

This microorganism is capable of adopting to its environment due to diversity in its structure and functioning. When exposed to antibiotics the bacteria survive by developing antibiotic resistance through various mechanical and chemical processes, giving it a lead against the antibiotics acting on it. The development of antibiotic resistance has been testified at some point of host colonization of CF patients, wherein *P. aeruginosa* strains increase and accumulate resistance for the duration of antimicrobial therapy. Research has suggested a robust association between multiple usages of ciprofloxacin with the accelerated incidence of ciprofloxacin-resistant strains. Therefore, some other element connected with the growth of MDR *P. aeruginosa* is related to the frequent use of antimicrobial agents. The development of this resistance can be a result of the consequence of a mutational instance or purchase of resistance gene through direct gene transfer which may rise at some point during antibiotic therapy, mutational events may lead to overexpression of endogenous beta-lactamases or efflux pump, and expression of specific porins.

Bacterial cell wall peptidoglycan synthesis is inhibited by Beta-lactam antibiotics. This group of drugs like penicillin, cephalosporin, carbapenem, and monobactam are usually effective against *P. aeruginosa*. Among these drugs, piperacillin and ticarcillin (penicillins), ceftazidime (third-generation cephalosporin), cefepime (fourth-generation cephalosporin), aztreonam (monobactam), imipenem, meropenem, and doripenem (carbapenems) are efficacious beta-lactam that is frequently used in of *Pseudomonas aeruginosa* treatment. B-lactamase enzymes work as key factors in developing resistance to B-lactam antibiotics. These develop resistance by breaking the amide bond of the B-lactam ring which results in demolishing the effect of the antibiotic therapy. The action of the drug is countered either by the expression of endogenous B-lactamases or by the expression of acquired beta-lactamases. Characterized by their substrate specificity hundreds of B-lactamases have been identified so far. Amber's molecular classification has put forward four major classes of B-lactamases found in *P. aeruginosa*: A-D. Classes A, C, and D catalyze the serine residue which inactivates the B-lactams whereas class B or Metallo-B-lactamases (MBLs) utilizes zinc.

P. aeruginosa is evidently vulnerable to carboxypenicillins, ceftazidime, and aztreonam however it could attain resistance through mutation in a gene that may result in hyper-manufacturing of AmpC beta-lactamase.

P. aeruginosa produces an inducible chromosome encoded AmpC beta-lactamase (cephalosporins) that belongs to molecular class C, based totally on Ambler and the first functional institution. Usually, the enzyme is produced in low quantities ('low-level' expression) and determines resistance to aminopenicillins and most of the early cephalosporins.

Fluoroquinolones develop resistance when mutation occurs within the gene encoding for DNA gyrase or topoisomerase or by active transport of drug out of the cell. Mutation of topoisomerase may occur in gyrA/gyrB genes at the active site of the enzyme, which is the determinative region motif for quinolone-resistance. This leads towards altered amino acid sequences of the A and B subunit, and hereafter a modified topoisomerase II with low binding affinity to quinolone molecules are formed. Alterations of topoisomerase IV occur as a result of point mutations in ParC and ParE genes encoding ParC and ParE enzyme subunits, individually. Another mode for fluoroquinolone resistance involves the overexpression of efflux [13].

4.6 How can we prevent infection?

Patients and care providers have to:

Preserve their palms so that they do not get contaminated and also do not spread germs that can cause infection.

Clean hands with soap and water or use a hand sanitizer containing alcohol prior to touching any wounds or handling any medical device.

Remind healthcare providers and caregivers to clean their palms before touching the patient or dealing with scientific devices.

Allow healthcare group of workers to clean their room daily whilst in a healthcare place.

Healthcare vendors need to be careful to suggest contamination control, which includes hand and environmental hygiene (e.g., cleansing of affected patient rooms and shared devices) to reduce the chance of spreading those germs to other patients.

Healthcare centers need to have appropriate water control plans that help ensure water quality and reduce the risk of exposure to potentially harmful germs like *P. aeruginosa* [6].

5. Conclusion

Healthcare-associated infections are worth preventing in terms of benefits in morbidity, mortality, duration of hospital stays, and cost. Educational interventions promoting good hygiene and aseptic techniques have a great effect on the control. Constant vigilance and attention by healthcare providers can be efficient to maintain their personal hygiene and the devices associated with the treatment, which contributes to lessening the occurrence of HAI.

Author details

Anika Farzin[1]*, Md. Mizanur Rahman[2] and Ferdows Ara Mollika[3]

1 Department of Microbiology, Eastern Medical College, Dhaka, Bangladesh

2 Department of Microbiology, Sylhet MAG Osmani Medical College, Sylhet, Bangladesh

3 Department of Microbiology, Tairunnessa Memorial Medical College, Gazipur, Bangladesh

*Address all correspondence to: anika.farzin11@gmail.com

References

[1] Labovska S. *Pseudomonas aeruginosa* as a cause of nosocomial infection. In: *Pseudomonas aeruginosa*—Biofilm Formation, Infections and Treatments. Croatia: IntechOpen Open Access Books; 2021. DOI: 10.5772/intechopen.95908

[2] National Nosocomial Infections Surveillance (NNIS). System Report, data summary from January 1992 through June 2004, issued October 2004. American Journal of Infectious Control. 2004;**32**:470-485. DOI: 10.1016/j.ajic.2004.10.001

[3] Gray J. Surveillance of Hospital Infection. Journal of Hospital Infection. 2016. Available from: www.journal.elsevier.com/journal-of-hospital-infection

[4] Spagnolo AM, Sartini M, Cristina ML. *Pseudomonas aeruginosa* in the healthcare facility setting. Reviews in Medical Microbiology. 2021;**32**(3):169-175. DOI: 10.1097/MRM.0000000000000271

[5] WHO. Prevention of Hospital-Acquired Infections. A practical guide. In: Ducel G, Fabry J, Nicolle L, editors. World Health Organization Department of Communicable Disease, Surveillance and Response. 2nd ed. Geneva, Switzerland: WHO; 2002

[6] Mulligan JA, Rushby JF, Adam T, John's B, Mills A. Unit Costs of Health Care Inputs in Low and Middle-Income Regions. Available from: https://www.researchgate.net/publication/228582375_Unit_costs_of_health_care_inputs_in_low_and_middle_income_regions [Accessed: September 23, 2022]

[7] Centers for Disease Control and Prevention (CDC). Healthcare-Associated Infections (HAI), *Pseudomonas aeruginosa* in Healthcare Settings. Atlanta: CDC; 2019

[8] Murray PR, Rosenthal KS, Pfaller MA. Medical Microbiology. 8th ed. Philadelphia: Elsevier; 2016. pp. 273-277

[9] Carroll KC, Hobden JA, Miller S, Morse SA, Meitzner TA, Detrick B. Jawetz, Melnick & Adelberg's Medical Microbiology. 28th ed. China: McGraw Hill Education; 2019

[10] Barat LF, Ferrer M, Rosa FD, Cabarrus A, Esperatti M, Terraneo S, et al. Intensive care unit-acquired pneumonia due to *Pseudomonas aeruginosa* with and without multidrug resistance. The Journal of Infection. 2017;**74**(2):142-152. DOI: 10.1016/j.jinf.2016.11.008

[11] Center for Disease Control and Prevention (CDC). Multidrug Resistant *Pseudomonas aeruginosa*: CDC'S 2019 Antibiotic Threat Report. Atlanta: CDC; 2019

[12] Islam TA, Shamsuzzaman SM, Nehar N, Fardows J. Prevalence and antibiogram of microbial agents causing nosocomial urinary tract infection in surgical ward of Dhaka Medical College Hospital. Journal of Enam Medical College. 2016;**6**(2):75-79. DOI: 10.3329/jemc.v6i2.27761

[13] Pachori P, Gothawal R, Gandhi P. Emergence of antibiotic resistance *Pseudomonas aeruginosa* in intensive care unit; a critical review. Genes & Diseases. 2019;**6**:109-119. Available from: http://ees.elsevier.com/gendis/default.asp

[14] Centers for Disease Control and Prevention (CDC). Antimicrobial Resistance. Atlanta: CDC; 2022

Chapter 4

Optimizing Shelf-life of *Pseudomonas fluorescens* after Freeze Drying

Nirmal Chandra Barman, Mohammad Sharif Sarker, Mahir Ahmed, Zahur Ahmed and Sankar Ramachandran

Abstract

The excess use of chemical fertilizers diminishes soil fertility and yield from crops gradually. To regain and enhance our soil nutrients to get more yields, it is mandatory to rely on soil microbes. Some beneficial microbes' termed as bio-fertilizers especially rhizosphere bacteria have well contribution in increasing plant growth, and yield without any toxins. It is a very natural process of interaction between plant and some microbes to increase the assimilation of nutrients and it helps plants to enhance better production. In this research, we showed the use of microbes especially *Pseudomonas fluorescens* survival in the soil. We applied different carriers such as dextrose, talc, and peat with freeze-dried *P. fluorescens* and studied the shelf-life of the *P. fluorescens.* Among different carrier's peat with centrifuged cell suspension survived up to 60 days with significant CFU's 2×10^7/gm CFU's, our research will be a prospective to make new formulations and to increase the shelf-life and survival rate of soil microbes.

Keywords: *Pseudomonas fluorescens*, centrifugation, lyophilization, dextrose formulation, peat formulation, talc formulation, shelf-life

1. Introduction

The world observes so many challenges in the food and agriculture sector especially in consuming quality and nontoxic foods. The excessive use of chemicals and chemical fertilizers is to enhance food production causes soil crust, infertile, acidification, and heavy metals accumulation in soil. As a consequence, long-term effects have been created in the land such as soil erosion and environmental deterioration. The agricultural yield, production, and food security are being globally hampered hugely by the climate change [1]. Therefore, the world needs a natural sustainable solution that can encourage environment-friendly approach with boosting up sound ecology, biodiversity, and resilient form of agriculture [2]. Currently, agro-experts and researchers are looking for the solution with the implementation of local crops cultivation, varieties, and indigenous strategies. Some beneficial microbes' termed as bio-fertilizers especially rhizosphere bacteria have well contribution in increasing plant growth, and yield without any toxins. It is a very natural process of interaction between plants and some microbes increase the assimilation of nutrients and it helps plants to enhance better production.

Bio-fertilizers are microbial-based fertilizers that provide essential NPK (Nitrogen, Phosphorous, and Potassium) nutrients to plants and reestablish the soil integrity. Various microorganisms are utilized as bio-fertilizers around the world such as *Rhizobium, Bacillus spp., Mycorrhizae, Azotobacter,* and *Pseudomonas* [3, 4].

Among other microbes, *Pseudomonas spp* is one of the notable bacteria which acts as bio-fertilizer as well as biocontrol agents. They play a dual role on crops; together they improve the plant growth, production, and resist the growth of pathogenic microorganisms. Alongside, they promote the establishment of other rhizosphere bacteria with roots [5]. Off all species of different *Pseudomonas spp,* researchers found *P. fluorescens* solubilize the insoluble phosphate source, mobilize them and increase the availability of phosphorus to plants [6]. They also enhance to assimilate nitrogen and produce phytohormones that encourage the vegetative growth of plants. *P. fluorescens* is well-known bio-fertilizer in agro-science, which has the ability to stimulate the plants growth that lives in contact with roots [7]. Researchers also proved that *P. fluorescens* resist the growth of pathogenic microorganisms of crop plants.

Using microbes as bio-fertilizers has a challenge for farmers due to their viability and survival in the soil. Many researchers investigated the viability and survival of the microbes which is used in organic farming. The researchers studied on various liquid- and solid-based microbial formulations as bio-fertilizers and their sustainability with significant results. Previous study showed that *P. fluorescens* bacteria can survive up to 6 months when glycerol was given in liquid formulation and it maintained the huge number of cells and viability well [8]. It was also found adding glycerol in nutrient broth media increased the survival of *Pseudomonas* cells which later showed efficacy against *Fusarium oxysporum* f. sp. *cubense* and *Helicotylenchus multicinctus* at multiple banana plantations [9]. A liquid formulation using humic acid for *P. fluorescens* has been developed which showed long shelf-life. The mixed formulation of humic acid along with *P. fluorescens* was used for crop protection and enhanced production [10].

The disadvantages of liquid bio-fertilizers are: it easily evaporates and diminishes their mode of action by sunlight. Solid bio-fertilizers (microbes with organic carrier materials) stabilize in sunlight and the chance of evaporation is less. Currently, there are different carriers such as peat, talc, corn flour, alginates, vermicomposte, and other organic materials. The solid bio-fertilizers are made by adding liquid colonies on to carrier. The difficulties of these carriers are they possess high moisture and humidity which causes microbial cell death.

Freeze drying or lyophilization of microbes and mixing with carrier may resolve this problem. So, here our research is intended to detect appropriate carrier particle and suitable technique to enhance the viability and sustainability of the microbes and to increase the shelf-life.

Although, *P. fluorescens* works very well as bio-fertilizers and has a good role in sustainable agriculture management, but the formulation and carrier materials are not fixed for long-time survival for solid formulation. Therefore, our focus also includes *P. fluorescens* enhanced production (more CFUs), to make suitable formulation and viability checking.

2. Materials and methods

Tryptone Type-1(Casitose Type-I) from Hi-Media laboratories, India; Yeast extract powder from Hi-Media laboratories, India; Sodium chloride from Merck, India; Dimethyl sulfoxide (DMSO) from Daejung, South Korea; skim milk powder (DANO) from Bangladesh; Agar powder from Merck, India; Dextrose from Hi-Media laboratories, India. And all other culture media were procured from Hi-Media laboratories, India and the reagents and chemicals from Merck, USA.

2.1 Isolation of *Pseudomonas fluorescens*

Soil sample was collected from rhizosphere of maize plant roots and brought to the lab. Then cetrimide agar media was prepared (pH 7.0 ± 0.2), autoclaved, poured, and 1.0 g soil was diluted in 0.9% saline solution. Diluted soil samples were spread on cetrimide agar plate and incubated for 72 hours. Vigorous growth colonies were picked up and subculture on LB plate. Light yellowish colonies were selected for the further process [11]. The strains were confirmed further by biochemical and 16 s rRNA genomic analysis.

2.2 Media preparation

Luria-Bertani (LB) a common media for microbial growth in the laboratory was used. The composition of media (g/l) is Tryptone 10 g, Yeast extract powder 5 g, and NaCl 10 g. pH was adjusted to 7.0 ± 0.2. LB Media was sterilized at 121°C, 15p si for 15 min. *P. fluorescens* strain was grown at 30°C in LB broth media [12–14] and was also maintained in paraffin at -86°C for further process.

2.3 Mother inoculum preparation

P. fluorescens (FLU-L) strain was routinely maintained in Luria-Bertani (LB) media. The broth media is used to prepare the mother inoculum for lab-scale production [14]. The 2% of previous cells were used as inoculum for lab-scale production. QC checked and recorded the CFUs.

2.4 PBS buffer (1X solution)

PBS buffer was prepared with the following composition (**Table 1**) and adjusted the pH to 7.0 Autoclaved the PBS buffer solution and cooled and stored at 4°C [15].

2.5 Lab-scale production

LB broth media was used for lab-scale production of inoculum [16]. We prepared 5 L for lab-scale production media, autoclaved at 121°C, 15 psi for 15 min, cooled and added 2% mother inoculum, and incubated at 30°C, 160 rpm, and 22 h at the shaker incubator. The media pH was kept at 7.0 ± 0.2. The growth of bacteria and CFU was analyzed.

2.6 Centrifugation

After incubation, the 5 L bacteria culture was aseptically transferred to centrifuge tubes for centrifugation. The centrifuge (CENLEE, Model: CFLR8, made in China) was

Reagents	g/l
NaCl	8 g
KCl	0.2 g
Na2HPO4	1.44 g
KH2PO4	0.24 g
Water	1 L
pH	7.0

Table 1.
Composition of the PBS buffer.

precooled at 4°C and centrifugation was done at 4000 rpm for 5 mins [9]. After centrifugation, the pellet was collected and the supernatant was discarded. Repeated the same procedure twice with PBS buffer. The QC analysis was done by checking CFU of the pellet.

2.7 Pre-freezing

Different ingredients can be used as cryoprotectant and give good survival of bacteria. Here, we used 10% Skim milk and 1% Dimethyl sulfoxide (DMSO) as cryoprotectant [17]. The cell pellets and cryoprotectant ratio was maintained equally and mixed gently. Subsequently, pre-freezing of mixed cell pellets was initiated at 4°C for 2 hrs in the refrigerator, 4 hrs in-20°C and at last it was kept overnight at -86°C.

2.8 Lyophilization

Freeze drying process was completed in a lyophilizer machine [18]. The pre-freeze cell pellet from the −86°C was freeze dried in the lyophilizer (Bioevopeak, model: LYO80V-2S), made in China). Running time can be variable on the basis of sample amount. The freeze drying was done for 50 hrs. After 50 hours, the dry pellets were collected and stored at 4°C for further process. The QC analysis was done by checking CFU of the dried cells.

2.9 Formulation

Formulations of the lyophilized dried (LP) cells were mixed with different carrier particles such as dextrose, peat soil, and talc and kept it for shelf-life analysis. The CFU of the formulated products was checked initially and frequently with time intervals.

2.10 Dextrose formulation

The 5% and 10% freeze-dried microbes were mixed with 95% and 90% dextrose, respectively, and kept it for shelf-life analysis at room temperature.

2.11 Peat formulation

We used two different methods, that is, solid and liquid to make peat formulations. For solid formulations, we mixed 5% lyophilized microbes and 95% peat (with less than 5% moisture) and kept it at room temperature for shelf-life analysis. We also

5% Dextrose	0.5 g
1% Tryptone	0.1 g
1% PVP	0.1 g
Pellet	10 g
Water	100 ml

Table 2.
Pellet solution preparation.

made liquid formulations by using cell pellets suspended in the following solution (**Table 2**) and mixed with the peat powder. Then 20 ml liquid pellet suspension was mixed with 70 g peat soil fine powder. Initially, the peat was grinded into fine powder, and pH was adjusted to pH 7 with calcium carbonate (CaCO3).

2.12 Talc formulation

The 5% and 10% freeze-dried microbes were mixed with 95% and 90% talc, respectively, and kept it for shelf-life analysis at room temperature.

2.13 Shelf-life analysis

After different formulations, we analyzed the finished products quality and its life span. We assayed our products periodically at 0, 7, 14, 30, 60, 90, 120, 150, and 180 days interval.

3. Results

3.1 *Pseudomonas* spp. morphology

Colony morphology and microscopic analysis of *P. fluorescens* were observed and recorded as shown below (**Figure 1**). *P. fluorescens* showed pale yellowish cream color with smooth moderate shape colonies. In microscopic, it showed a small rod shape, gram-negative bacteria.

3.2 Mother inoculum

P. fluorescens (FLU-L) showed yellowish turbid growth after 22 hrs incubation. We observed that the mother culture CFU was 1.0×10^8/ml (**Figure 2**).

3.3 Lab-scale production

The 5 L *P. fluorescens* grown flasks in LB media were observed for CFU and showed 1.0×10^8/ml. Culture formed a yellowish appearance in LB broth (**Figure 3**).

3.4 Centrifugation

After centrifugation, the yield of the cells was 0.7–0.8% pellet from 5 L LB broth. Pellets were pinkish in color with smooth cream (**Figure 4**). The pellets CFUs showed 1.0×10^{10}/ml.

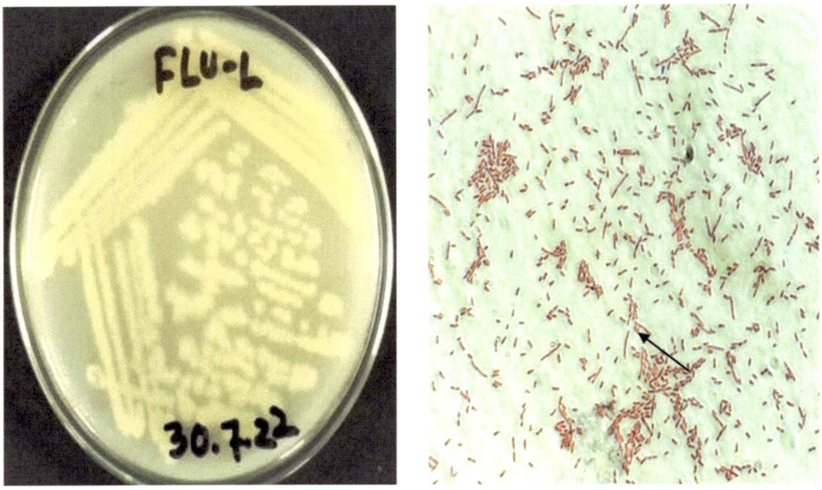

(a) (b)

Figure 1.
a) Pseudomonas fluorescens *colonies in LB plate; b)* Pseudomonas fluorescens *showed gram-negative rod shape bacilli in gram staining.*

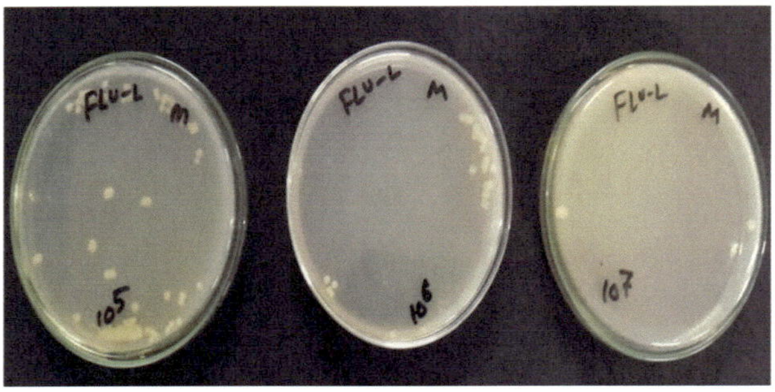

Figure 2.
CFU of mother inoculum.

3.5 Lyophilization

The lyophilized microbial powder (LP) was light brownish color and it showed 6% moisture during harvesting. The CFU of LP was recorded as 1.0×10^9 (**Figure 5**).

3.6 Formulation

We analyzed dextrose formulations in two configurations, 5% and 10%. Both compositions showed 4.0×10^7/gm (**Figure 6**) and 5.0×10^7/gm (**Figure 7**), respectively, on the formulation day (**Tables 3** and **4**).

3.7 Solid peat formulation

The mixture of 5% LP and 95% peat was also resembled dextrose results. It showed 4.0×10^7 /gm CFU's (**Table 5** and **Figure 8**).

(a)

(b)

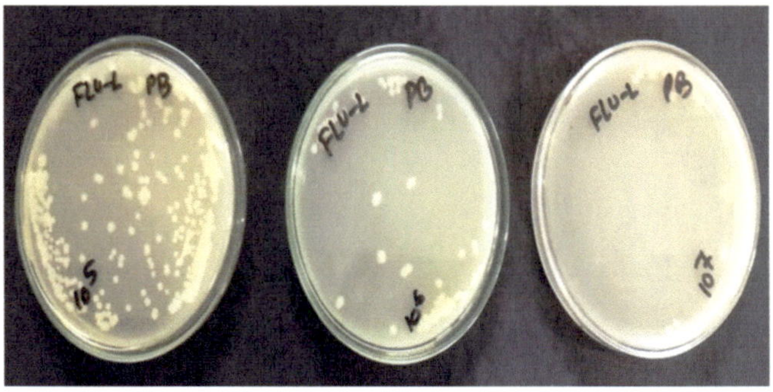

(c)

Figure 3.
a) Broth culture in flask; b) broth culture in centrifuged tube; c) CFU of production broth.

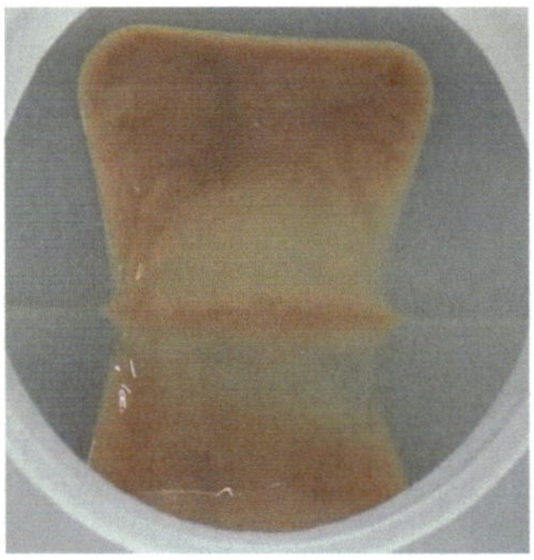

(a)

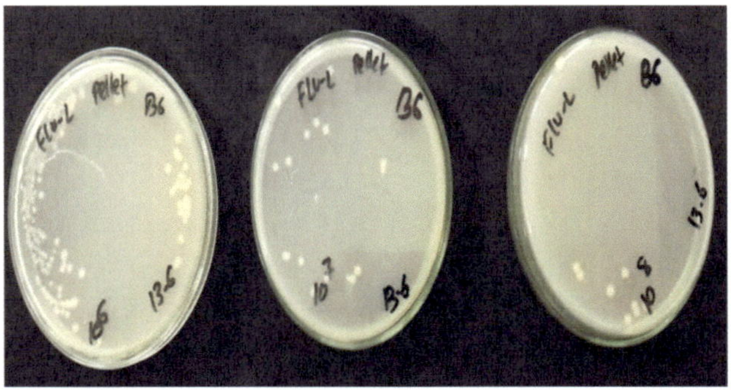

(b)

Figure 4.
(a) Picture of pellet; b) CFU of pellet.

3.8 Talc formulation

Our results showed that the mixture of LP and talc as a carrier was void. There was no CFU's recorded even after formulation (**Table 6**).

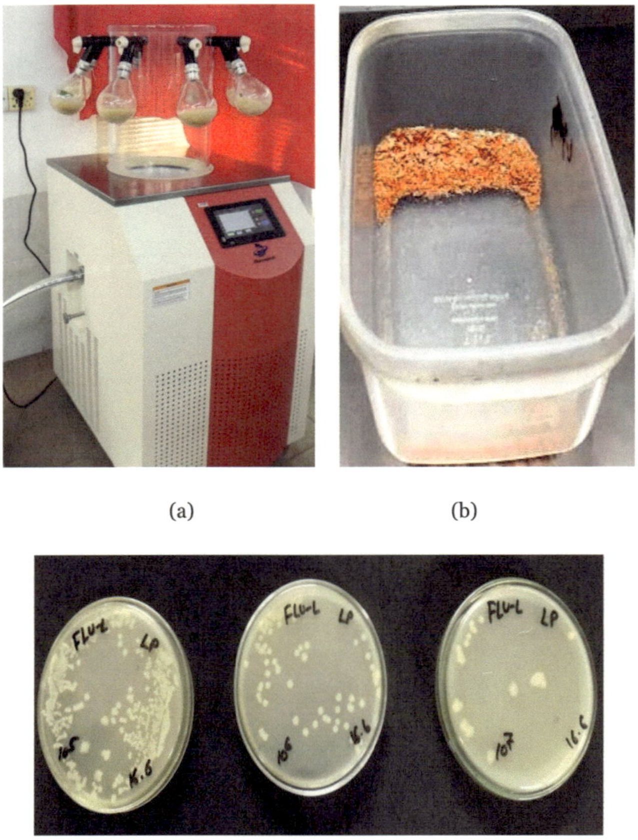

(a) (b)

(c)

Figure 5.
(a) Lyophilization is running; (b) Lyophilized powder (LP); (c) CFU of LP.

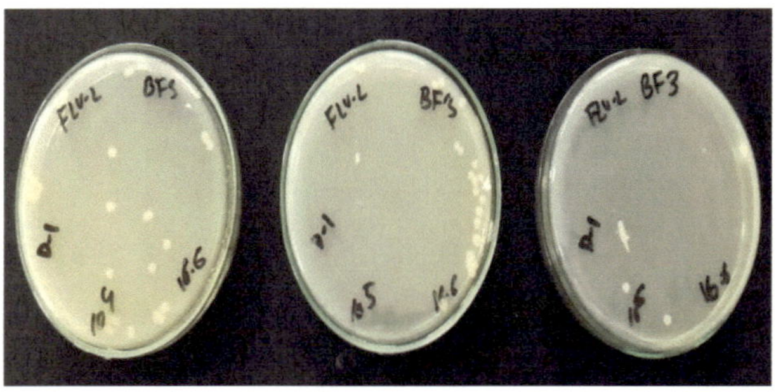

Figure 6.
Colonies in LB plate from 95% dextrose formulation.

3.9 Liquid peat formulation (with pellet suspension)

The pellets mixed with suspension formulations and added with peat showed us increasing CFU's 5.0×108/gm (**Table 7** and **Figure 9**) than other formulations on the formulation day.

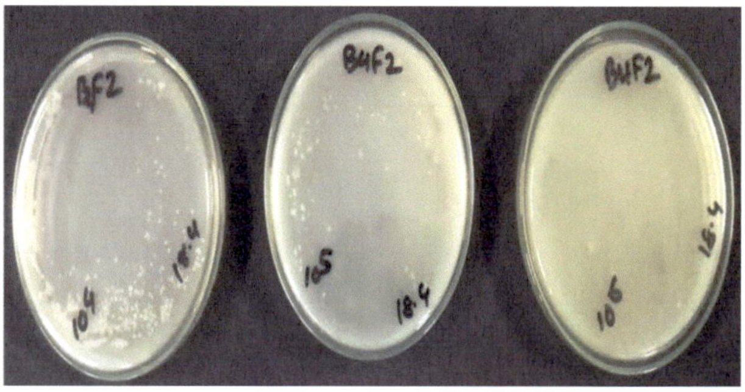

Figure 7.
Colonies in LB plate from 90% dextrose formulation.

Formulation	Days	pH	Moisture	CFU/gm
5% LP + 95% dextrose	0	≈7.00	7.10	4.0×10^7

Table 3.
Dextrose formulation.

Formulation	Days	pH	Moisture	CFU/gm
10% LP + 90% dextrose	0	≈7.00	6.98	5.0×10^7

Table 4.
Dextrose formulation.

Formulation	Days	pH	Moisture	CFU/gm
5% LP +95% peat	0	≈7.0	6.11	4.0×10^7

Table 5.
Solid peat formulation.

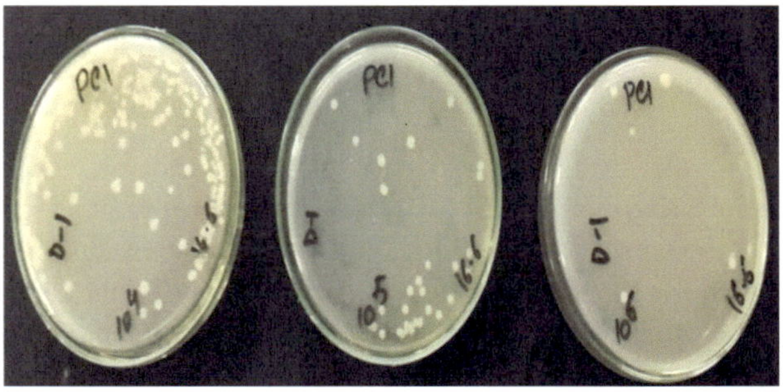

Figure 8.
Colonies in LB plate from solid peat formulation.

3.10 Shelf-life analysis

In this research, we also revealed the shelf-life of LP microbes with different formulations such as dextrose, peat, and talc. Peat with centrifuged liquid cell

Formulation	Days	pH	Moisture	CFU/gm
5% LP +95% Talc	0	≈7.0	6.99	00.00

Table 6.
Talc formulation.

Formulation	Days	pH	Moisture	CFU/gm
70% peat+30% CaCO3 + 20 ml pellet suspension	0	≈7.0	17.48	5.0×10^8

Table 7.
Peat formulation with pellet suspension.

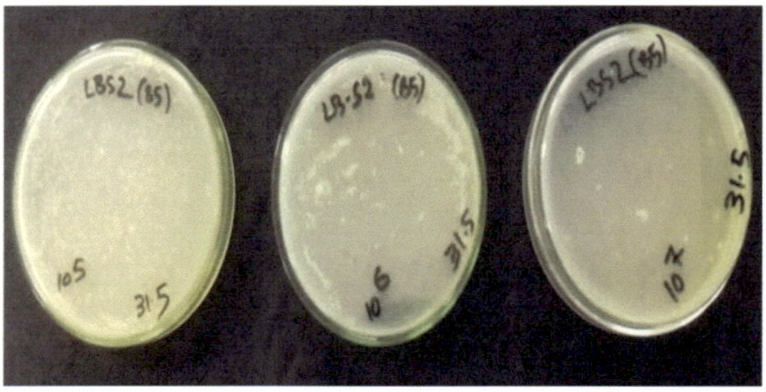

Figure 9.
Colonies in LB plate from liquid peat formulation.

Formulation	Days	pH	Moisture	CFU/gm
5% LP + 95% dextrose	0	≈7.00	7.10	4.0×10^7
	7	≈7.00	4.83	00.00
	14	≈7.00	6.91	00.00

Table 8.
Dextrose formulation shelf-life.

suspension showed significant shelf-life up to 30 days. It showed 1.0×10^7/gm CFU's. Rather the 5% (**Table 8** and **Figure 10**) and 10% LP (**Table 9** and **Figure 11**) and 95% and 90% dextrose showed no colonies on 07 days' shelf-life analysis. The LP and the peat mixture survived for 14 days with 4.0×10^4/gm CFU's (**Table 10** and **Figure 12**). Previously, we showed that the talc formulations were not survived even a day and this proves that centrifuged pellet liquid suspension mixed with peat gives us the 60-day survival significant CFU's (**Table 11** and **Figure 13**).

4. Discussion

The inclusive results of our experiments showed that *P. fluorescens* could survive initially with dextrose, solid peat, and talc carriers at room temperature. But *P. fluorescens* was not viable in dextrose carrier up to 1 week. Peat formulation was found as a

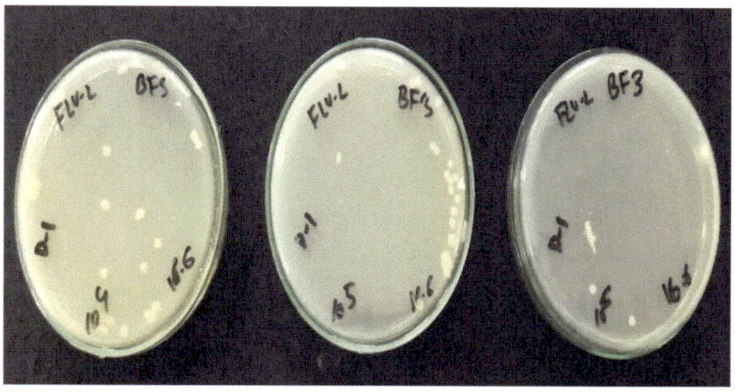

(a)

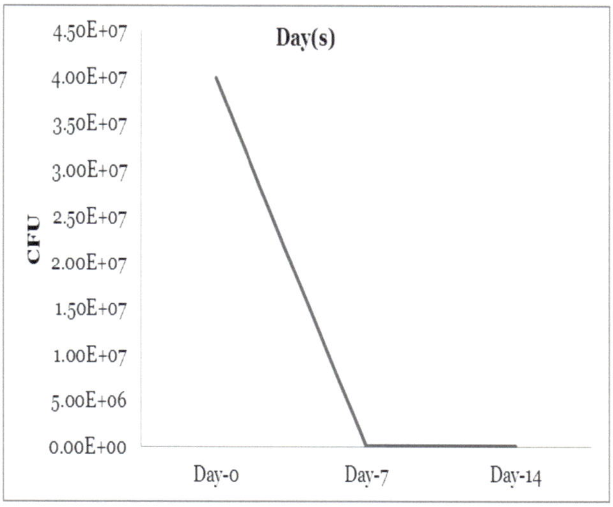

(b)

Figure 10.
Shelf-life colonies in LB plate.

Formulation	Days	pH	Moisture	CFU/gm
10% LP + 90% dextrose	0	≈7.00	6.98	5.0×10^7
	7	≈7.00	5.08	00.00
	14	≈7.00	8.44	00.00

Table 9.
Dextrose formulation shelf-life.

good carrier for formulating product of *P. fluorescens*. When LP was mixed with peat powder, it was observed cell number 4×10^4 per gram till 14 days. The best significant result was observed when peat was mixed with centrifuged cell pellet suspension; we observed 2×10^7 CFUs per gram surviving more than 60 days. Peat is very common used carrier material globally and has high content of organic matter, water holding capacity, and simply obtainable [19]. In previous study, it was found that peat showed good efficacy in formulation, and help colonization in plant roots when peat was used for bio formulation of *P. fluorescens* [20]. Nakkeeran et al. also observed similar potential results of peat-based formulation for *P. chlororaphis* shelf-life [21]. Therefore, we also reveal that peat with less moisture is the suitable carrier for centrifuged

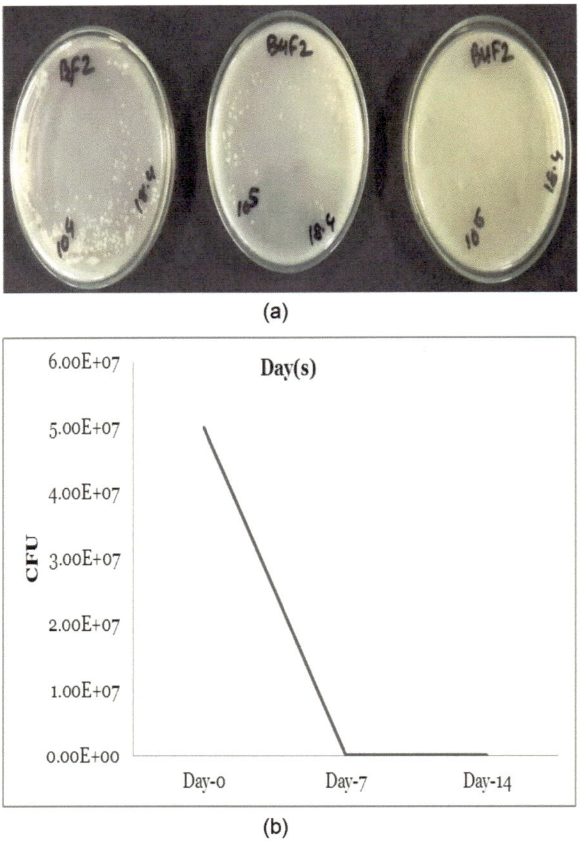

Figure 11.
Shelf-life colonies of 10% LP + 90% dextrose in LB plate.

Formulation	Days	pH	Moisture	CFU/gm
5% LP +95% peat	0	≈7.0	6.11	4.0×10^7
	07	≈7.0	6.29	2.0×10^5
	14	≈7.0	7.75	4.0×10^4
	30	≈7.0	5.83	00.00

Table 10.
Solid peat formulation shelf-life.

cells suspension. Our research will be prospective to make new formulations and to increase the shelf-life and survival rate of soil microbes especially *P. fluorescens*.

5. Conclusion

From our results, we can conclude that centrifuged cells suspended in peat survived for long duration. Applying peat with suspended *P. fluorescens* may not hamper

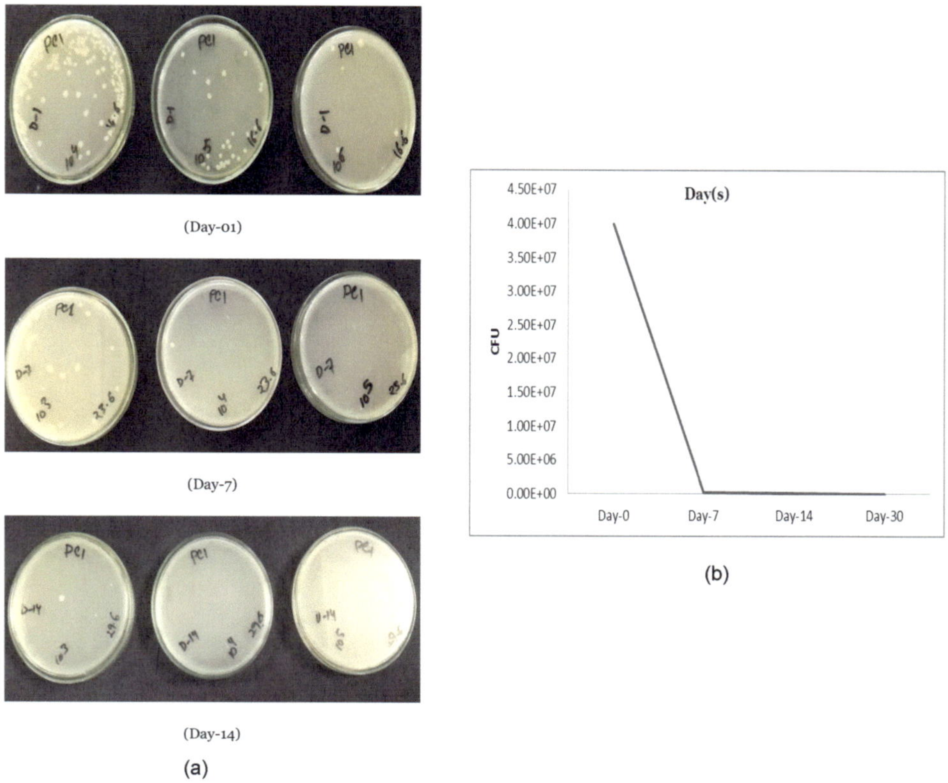

Figure 12.
Shelf-life colonies of 95% peat+5% LP in LB plate.

Formulation	Days	pH	Moisture	CFU/gm
70% peat+30% $CaCO3$ + 20 ml pellet suspension	0	≈7.0	17.48	5.0×10^8
	7	≈7.0	18.60	6.0×10^7
	14	≈7.0	20.22	4.0×10^7
	30	≈7.0	22.13	1.0×10^7
	60	≈7.0	18.96	2 × 107

Table 11.
Peat formulation with pellet suspension shelf-life.

soil fertility, since peat is a universal bio carrier and sustains with all soil conditions. Even though we tried different carriers such as dextrose and talc, peat termed to be a suitable carrier for *P. fluorescens.*

In conclusion, although different types of carrier can be mixed with *P. fluorescens* for bioformulation, peat-based formulation showed good viability at room temperature.

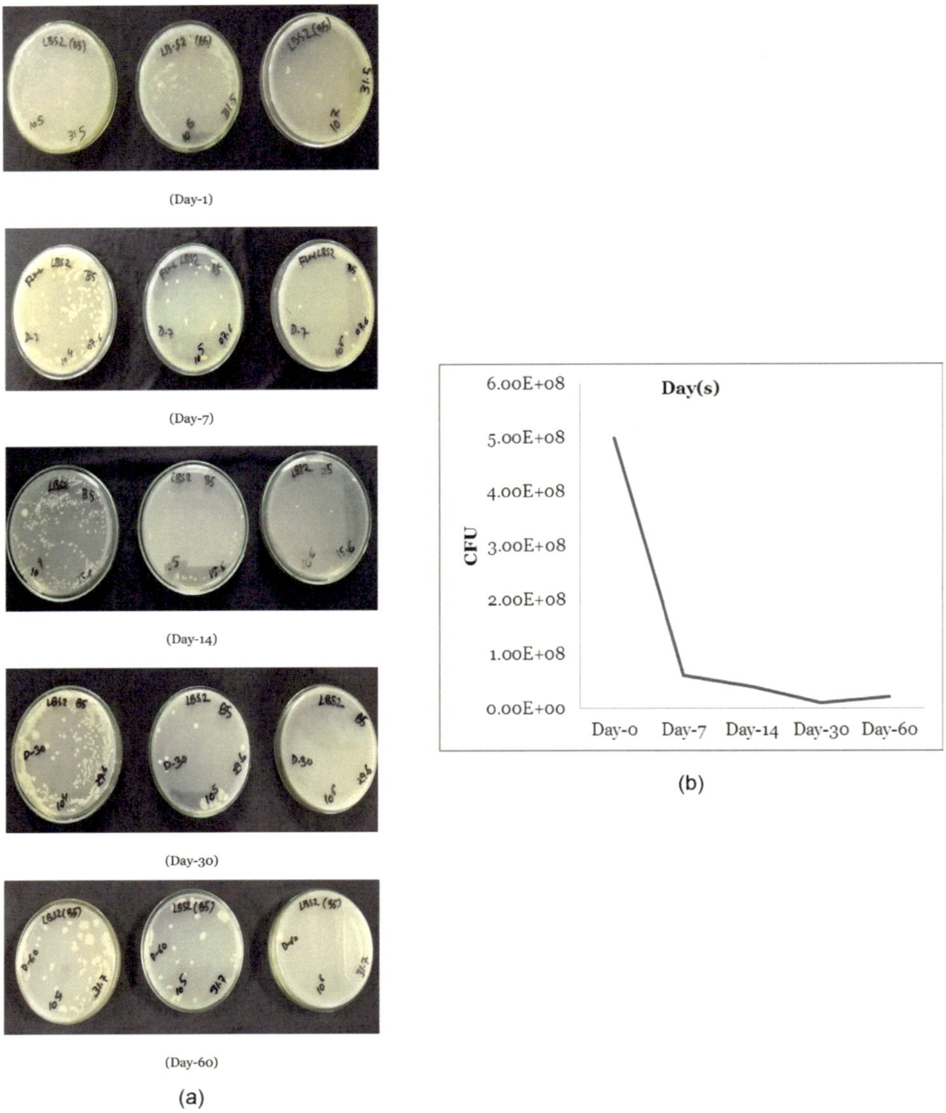

Figure 13.
Shelf-life of 70% peat+30% $CaCO_3$ + 20 ml pellet suspension in LB plate.

Acknowledgements

We express our gratitude and thanks to our Managing Director and Director, Apex bio-fertilizers and bio-pesticides Ltd. for supporting and funding this research. We also extend our thanks to our product development lab members and their associates.

Author details

Nirmal Chandra Barman, Mohammad Sharif Sarker, Mahir Ahmed, Zahur Ahmed and Sankar Ramachandran*
Product Development Lab, Department of Biotechnology, Apex Bio-Fertilizers and Bio-Pesticides Ltd., Dhaka, Bangladesh

*Address all correspondence to: sankar@apexbiofertilizer.com

References

[1] Suprasanna P. Plant abiotic stress tolerance: Insights into resilience build-up. Journal of Biosciences. 2020;**45**(1):1-8

[2] Santana-Fernández A, Beovides-García Y, Simó-González JE, Pérez-Peñaranda MC, López-Torres J, Rayas-Cabrera A, et al. Effect of a *pseudomonas fluorescens*-based biofertilizer on sweet potato yield components. Asian Journal of Applied Sciences. 2021;**9**(2)

[3] Martínez LR, Sánchez DC, Cuellar EE, González JS, Espinosa RR, Cuellar AE. Efecto de las micorrizas y bioplaguicidas sobre cultivares de raíces y tubérculos en un suelo pardo mullido carbonatado. Agricultura Tropical. 2015;**1**(1)

[4] Pérez-Pazos JV, Sánchez-Lopez DB. Characterization and effect of Azotobacter, Azospirillum and *Pseudomonas associated* with Ipomoea Batatas of Colombian Caribbean. Revista Colombiana de Biotecnología. 2017;**19**(2):35-46

[5] Imperiali N, Chiriboga X, Schlaeppi K, Fesselet M, Villacrés D, Jaffuel G, et al. Combined field inoculations of *Pseudomonas* bacteria, arbuscular mycorrhizal fungi, and entomopathogenic nematodes and their effects on wheat performance. Frontiers in Plant Science. 2017;**8**:1809

[6] Nieto P. *Pseudomonas*, Microorganismos de Biocontrol en Agricultura. Control Bío 2016. Available from: https://controlbio.es/es/blog/c/92_*pseudomonas*-microorganismos-de-biocontrol-en-agricultura.html. [Accessed: November 6, 2018]

[7] Madigan M, Martinko J. In: Pearson, editor. Brock Biology of Microorganisms. 15th ed. Harlow, UK: Pearson Educational Limited; 2019. p. 1056

[8] Manikandan R, Saravanakumar D, RajendranL,RaguchanderT,SamiyappanR. Standardization of liquid formulation of *Pseudomonas fluorescens* Pf1 for its efficacy against Fusarium wilt of tomato. Biological Control. 2010;**54**(2):83-89

[9] Selvaraj S, Ganeshamoorthi P, Anand T, Raguchander T, Seenivasan N, Samiyappan R. Evaluation of a liquid formulation of *Pseudomonas fluorescens* against *Fusarium oxysporum* f. sp. *cubense* and *Helicotylenchus multicinctus* in banana plantation. BioControl. 2014;**59**(3):345-355

[10] Pushpa A, Subhash PC, Reddy MAH. Development of liquid formulation for the dual purpose of crop protection and production. Journal of Environmental Research and Development. 2014;**8**(3):378

[11] Düyüncü D, Ulusoy S. *Pseudomonas fluorescens* isolation from green salads and antibiotic susceptibilities of isolates. Akademik Gıda. 2019;**17**(4):444-449

[12] Ramachandran S, Singh M, Mandal M. Purification of Azurin from *Pseudomonas aeuroginosa*. In: Chromatography-The Most Versatile Method of Chemical Analysis. Vol. 24. London, UK: InTechOpen; 2012. pp. 311-324

[13] Sankar R, Mandal M. Azurin synthesis from *Pseudomonas aeruginosa* MTCC 2453, properties, induction of reactive oxygen species, and p53 stimulated apoptosis in breast carcinoma cells. Journal of Cancer Science and Therapy. 2011;**3**(5):104-111

[14] Kirner S, Krauss S, Sury G, Lam ST, Ligon JM, van Pée KH. The non-haem

chloroperoxidase from *Pseudomonas fluorescens* and its relationship to pyrrolnitrin biosynthesis. Microbiology. 1996;**142**(8):2129-2135

[15] Faizal I, Dozen K, Hong CS, Kuroda A, Takiguchi N, Ohtake H, et al. Isolation and characterization of solvent-tolerant *Pseudomonas putida* strain T-57, and its application to biotransformation of toluene to cresol in a two-phase (organic-aqueous) system. Journal of Industrial Microbiology and Biotechnology. 2005;**32**(11-12):542-547

[16] Phosphate Buffered Saline (PBS): Composition and Preparation [Internet]. Share Biology. [cited 2023 Feb 23]. Available from: https://sharebiology.com/phosphate-buffered-saline-pbs-composition-and-preparation/

[17] Vesper SJ. Production of pili (fimbriae) by *Pseudomonas fluorescens* and correlation with attachment to corn roots. Applied and Environmental Microbiology. 1987;**53**(7):1397-1405

[18] Navarta LG, Calvo J, Posetto P, Benuzzi D, Sanz MI. Freeze-drying of a mixture of bacterium and yeast for application in postharvest control of pathogenic fungi. SN Applied Sciences. 2020;**2**(7):1-8

[19] Bhattacharyya C, Roy R, Tribedi P, Ghosh A, Ghosh A. Biofertilizers as substitute to commercial agrochemicals. InAgrochemicals detection, treatment and remediation. Butterworth-Heinemann. 2020;**1**:263-290

[20] Novinscak A, Filion M. Long term comparison of talc-and peat-based phytobeneficial *Pseudomonas fluorescens* and *Pseudomonas synxantha* bioformulations for promoting plant growth. Frontiers in Sustainable Food Systems. 2020;**4**:602911

[21] Nakkeeran S, Kavitha K, Mathiyazhagan S, Fernando WG, Chandrasekar G, Renukadevi P. Induced systemic resistance and plant growth promotion by *Pseudomonas chlororaphis* strain PA-23 and *Bacillus subtilis* strain CBE4 against rhizome rot of turmeric (*Curcuma longa* L.). Canadian Journal of Plant Pathology. 2004;**26**:417-418

Chapter 5

The Pangenome of *Pseudomonas aeruginosa*

Mauricio Corredor, Juan David Patiño-Salazar, Diana Carolina Castaño and Amalia Muñoz-Gómez

Abstract

This review summarizes the most important reports about *Pseudomonas aeruginosa* pangenome. Pan-genomics has tackled some fundamental concerns in pathogenic bacteria. PATRIC and other databases, store more than 9000 *P. aeruginosa* genomes. This data mining is an opportunity to develop discoveries related to antibiotic resistance, virulence, pathogenicity, fitness, and evolution, among others. Observing the different pangenomes of *P. aeruginosa*, it is concluded that this species has an open pangenome, and its accessory genome is larger than the central genome. HGT is one important source for *P. aeruginosa* genome. In recent years various authors developed *P. aeruginosa* pangenomes, from works with five genomes to more than 1300 genomes. This last work analyzed 54,272 genes, and they found a short and tiny core genome (only 665 genes). Other research with lesser strains or genomes identified a core genome bigger, almost 20% of the pangenome. Nevertheless, the total work proves that the accessory plus unique genome is larger than the core genome in *P. aeruginosa*.

Keywords: pangenome, pan-genome, *pseudomonas aeruginosa*, bacterias, antibiotic resistome

1. Introduction

Pan-genomics/pangenomics is an innovative tool to explain pan-genome/pangenome construction throughout the species, which is resolved with comparative genomics, among others. Pangenome is divided into two main classes: the core and accessory genomes. Thousands of unknown bacteria and microorganisms are exposed to natural and manufactured antibiotics, toxins, and harmful compounds, to the highest and lowest temperatures, extreme pHs, and other species competitors. Pangenomics is too a powerful approach to identifying those thousands of involved genes. Virulent genes, phenotypes, and environmental expressed genes from horizontal transference (HGT) derive from the core or accessory genome. These latest concepts are a challenge as a new point of view to face *Pseudomonas aeruginosa*: a new pathosystem, multidrug-resistant, and old human pathogen. To delve more specifically into pangenomics and pangenome, there have been published two remarked books recently [1, 2].

Different authors and web-published pangenomes' *P. aeruginosa* [3–7] https://pangenome.org/Pseudomonas_aeruginosa): However, other valuable databases

Tool	Webpage	Reference
Panseq	https://lfz.corefacility.ca/panseq/page/pan.html ; https://github.com/chadlaing/Panseq ; https://help.rc.ufl.edu/doc/Panseq ; https://opensourcelibs.com/lib/panseq	[13]
PanCGHweb	http://bamics2.cmbi.ru.nl/websoftware/pancgh/pancgh_start.php; https://maayanlab.cloud/datasets2tools/landing/tool/PanCGHweb	[14]
PGAT	http://nwrce.org/pgat/ ; http://tools.uwgenomics.org/pgat/index.html	[15]
Spine	http://vfsmspineagent.fsm.northwestern.edu/cgi-bin/spine.cgi http://spineagent.fsm.northwestern.edu/cgi-bin/spine.cgi	[16]
PGAweb	http://pgaweb.vlcc.cn/analyze ; stopped in English availability, only in Chinese language, http://www.cngrid.org/yhfw/zhsq1/#	[17]
PanX	https://pangenome.org/	[7]

Table 1.
List of some pangenomics tools. After publication, some addresses were inactivated. To obtain unavailable tools, please to contact the corresponding author from the references. More than 40 pangenomic tools are now available in online platforms or for local applications [12].

stocked *P. aeruginosa* genomes (https://www.pseudomonas.com/ or https://patricbrc.org/search/?and(keyword(Pseudomonas),keyword(aeruginosa)) with complete and draft 9954 genomes). These authors published different pangenomes using varying amounts of *P. aeruginosa* genomes: Sharma et al. [3] used 5 genomes, Fischer et al. [5] used 100 genomes, Ding et al. [7] used 153 genomes on the web page, Mosquera-Rendón et al. [4] used 181 genomes, and Freschi et al. [6] used 1311 genomes. We will see in the following subtitles, which will be the best amount of genomes to reach one open and close pangenome.

In the pangenome, the core or central genome is the total of genes common to all the examined and analyzed genomes from a genome pool of a given species. Likewise, this represents the genes present in the overall strains from one species. On the other hand, accessory, variable or flexible genome (for some authors, dispensable genome) represents the genes that do not present in all strains of one species [8], http://www.metagenomics.wiki/pdf/definition/pan-genomeb [9, 10]. Those terms are the key in pangenomics to reach a significant pangenome into some species. Moreover, the next sections will see the robustness degree or gene orthology analysis to have a core and accessory genome.

Concomitantly, multiple pangenomic tools were developed and tested over the last ten years. Of course, we can start from genomics annotation if we do not want to use the available databases since classical pangenomics use only annotated genes and orthology analysis. We will mention some of them because it is sure the high amount of pangenomics tools available today. Furthermore, now database strategy, currently developed to perform via a personal server [11], there are now online resources that allow to quickly build own pangenome analysis see **Table 1**.

2. What is a pangenome?

The pangenomic technique requires at least two or multiple genomes of bacteria, archaea, fungi, or any eukaryote. This tool can provide the broadest resolution of genetic variation, among them: pathogenicity islands, virulence genes, mobile elements, transposons, horizontal gene transfer, pathogenicity islands, orthologous

Terminology	Description
Pangenome or Pan-genome	It is the gene collection from a genomes group of species or strain group. Also, the complete gene set of all strains of a species, including genes present in all strains (core genome) plus genes present only in some strains of a species (variable or accessory genome).
Core or Central genome	It is the maximum of genes common to all the examined genomes of a given species. Likewise, represent the genes present in the overall strains from one species.
Accessory or Variable genome	The variable or accessory genome (flexible, dispensable genome) refers to genes not available in all strains of a species. These include genes present in two or more strains, or even unique genes from a single strain (for some authors). For example, genes for specific strain adaptation such as antibiotic resistance. Not uniques, but not in the core genome.
Unique genome	A set of genes present in only one strain, that is never shared with other strains. Singleton is the name assigned to this kind of gene. Some authors classify singletons as accessory genomes others prefer to differentiate them.
Open pangenome	It is known as the pangenome increasing when a new genome is added to the pangenome.The number of genes of the pangenome increases with the number of additionally sequenced strains.
Closed pangenome	Finished pangenome is when there is no change when new genomes are added, after some sequenced strains, additional strains do not provide new genes to the species pangenome.

Table 2.
Key terminology in the bioinformatics world [8]; http://www.metagenomics.wiki/pdf/definition/pan-genomeb [9, 10].

shared and syntheny, plasticity, evolution, and others. The development of pangenomics has promoted advances in many fields, like bioinformatics and computational biology, comparative genomics, molecular medicine, molecular epidemiology, agronomy and foods, and many more [18]. In the beginning, one advantage of pangenomics was that experimental data have shown for some species that new genes are being discovered even after sequencing several strain genomes [19]. Given that, the number of unique genes is vast, the pangenome of a bacterial species might be orders of magnitude larger than any single genome, as predicted by [20], more than 14 years ago.

The connection between the core and accessory genome put together the close and open pangenome inside the species. Open pangenome increases when new genomes are added, contrary to a close when a new genome added does not increase the size of the pangenome [10]. **Table 2** summarizes the terms applied to pangenome and pangenomics. It is worth mentioning that the term coined "unique genome", is related to solitary genes not shared among strains.

3. From pangenomics to pangenome

The term pan-genome, currently written today as pangenome was first used by Sigaux [21] in cancer research to describe a public database containing an assessment of genome and transcriptome alterations in types of tumors, tissues and experimental models. Later, Tettelin et al. [22], using bacterial genomes, defined a microbial pangenome as the combination of a core genome, carrying genes present in overall strains. And a dispensable genome (also described as a flexible or accessory genome) is composed of genes absent from one or more strains [23]. A generalization of such representation could contain not only the genes (transposons, promotors, other mobile elements such as HGT, microRNAs, etc., but also other variations present in the collection of genomes.

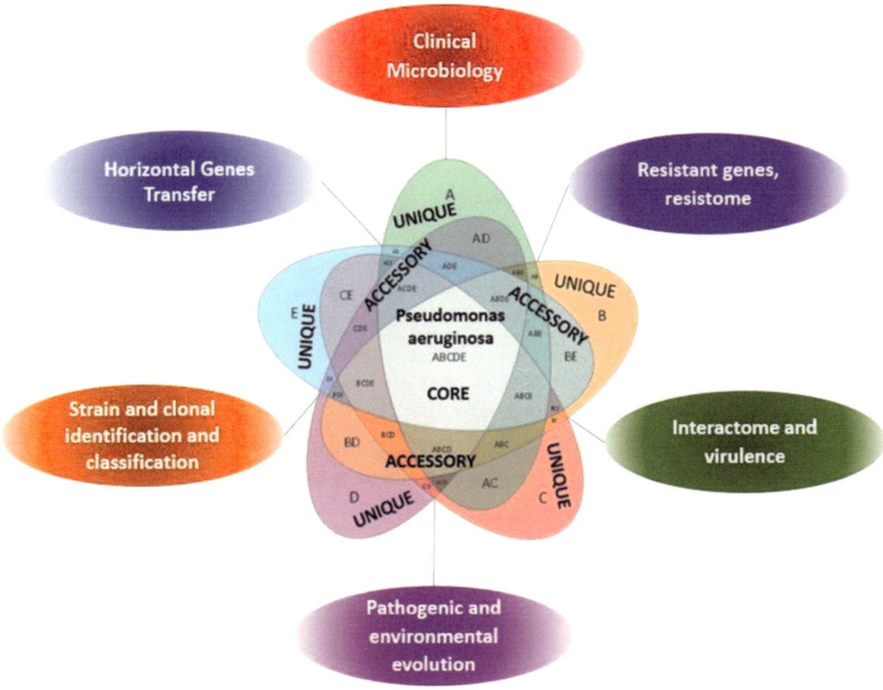

Figure 1.
Pangenomics tool, in the worldwide context. Pangenomics is useful for the treatment of patients in evolutionary studies. The HGT and virulent genes are investigated today with pangenomics to solve the MDR problem of P. aeruginosa.

The study of bacterial pangenome has many applications in clinical microbiology, also to study resistant genes or HGTs, virulence, and pathogenicity, and finally to classify species and to know its evolutions and clonal dispersion inside de environments see **Figure 1**. The pangenomics offer a wealth of information about human-associated bacterial species (12).

4. Could pangenome redefine bacteria species?

In his excellent chapter of Bobay [24], about the prokaryotic species concept, supported by the pangenome study, takes into account the approaches from Dobzhansky [25] and Mayr [26]. Speciation mechanism is involved in microorganisms, directly or indirectly, since the sustained interruption of gene flow between populations does not engage in sexual reproduction *stricto sensu*, escaping to the classic concept of species. Bobay [24] says that the definition of species has direct consequences regarding the definition of pangenome, and it is clear that 16S rRNA is not entirely cohesive, concluding and affirming: "studies focusing on the evolution of bacterial pangenomes should be based on rigorous species delimitation since the misclassification of a single genome can lead to dramatic overestimates or underestimates of the size of a species' pangenome".

Pangenomics is a fundamental tool for studying the entire repertoire of gene families in the genomes of pathogenic bacterial clades such as *P. aeruginosa.* This does not only provide the whole set of genes shared by this species, but also can be applied in interspecies differentiation analysis to mine species-specific genes to use a wealth of genome data [27]. Wang et al. [28] aimed at mining novel specific target gene sequences of *P. aeruginosa* based on the pangenome analysis and established high-specificity and high-sensitivity PCR and quantitative real-time PCR (qPCR) methods based on these targets. They used pangenome analysis to analyze the whole genomic sequences of 1,000 *P. aeruginosa* strains compared with other *Pseudomonas* species. A remarkable problem is the deficiency and mutations of some virulence factors in *P. aeruginosa* strains, which can result in false positives since existing pathogenic factors, may cause a potential threat of food poisoning (Baloyi et al. [29] cited by Wang et al. [28]).

The application of pangenomics in *P. aeruginosa* species with an open pangenome allows realizing that this species accepts external genes constantly and continuously from other strains and species, giving genetic richness and fitness to *P. aeruginosa*. Mutations over the genes and methylations in DNA and RNA gave a broad diversity to *P. aeruginosa*, expanding the variety and allowing success in diverse environments. Hilker et al. [30] compare some clonal genomes and indicate that the differential genetic repertoire of clones maintains a habitat-independent gradient of virulence in the *P. aeruginosa* population.

P. aeruginosa is excellently characterized in NCBI/taxonomy (https://www.ncbi.nlm.nih.gov/Taxonomy/Browser/wwwtax.cgi?id=287; [31, 32]. However, genomics and classical taxonomic definition (Pseudo, *false*), (monas, *unit*) are on different approaches; although the first is based on sequencing, the second is based on metabolism and 16S rRNA, it is significant to remember that 16S must be sequenced. Also, to define species is necessary a certain number of strains. Likely, classical taxonomy and pangenome have similarities: it needs a certain number of strains, when better, a large number of those ones. In summary, pangenome comes to put together genomics and classical taxonomy. The contribution of *P. aeruginosa* is remarkable because new tools must complement the information and research and never disperse the consensus. When Carl Wose and colleagues redefined [33–35] taxonomic classification based on 16S rRNA, the initial reluctance to classify was constant. Today the regular is classified with 16S rRNA. But when looking at certain strains and their metabolism was found that 16S rRNA also falls short of classification [36]. Precisely the pangenomics comes in response to the new inconsistencies of 16S rRNA [37, 38].

5. Core and accessory genome of *P. aeruginosa*

As previously mentioned, different *P. aeruginosa* pangenomes were developed by Sharma et al. [3], Fisher et al. [5], Ding et al. [7], Mosquera-Rendón et al. [4], Freschi et al., [6], with a different number of 5, 100, 153, 181, 1311 genomes, respectively. However, the works ranging from 100 to 181 genomes almost agreed on the size of core and accessory genomes. Mosquera-Rendón et al. [4] found approximately 15 % was the core genome (2503/16,820 genes) of the pangenome, while Ding et al. [7] in their web page for *P. aeruginosa*, 16,327 genes selected from 18,780 total genes are accessory genome, and 2453 genes are the core genome. Freschi et al.'s [6] pangenome of *P. aeruginosa* consists of 54,272 genes: 665 are core genes, 26,420 are accessory genes, and 27,187 are unique genes. This work has a robust cutoff bioinformatic to discriminate homologous genes, eliminating potential orthologous (**Figure 2**).

It will most likely be necessary to adjust the core and accessory genome measurements for *P. aeruginosa*, given the differences mentioned before. One facility to

Figure 2.
The P. aeruginosa pangenome from Ding et al. [7] in the interactive database https://pangenome.org/Pseudomonas_aeruginosa. This web developed with the panX tool allows comparing the pangenomes data of different pathogens and displays rapid results of pangenomics analysis.

improve this is that *P. aeruginosa* has become an influential species in many scenes, not just clinical or environmental. Therefore, the large number of strains coupled with improved bioinformatics tools, such as improving the complete sequencing of a large number of strains, will enhance the calculations between core and accessory genomes. Genomics and pangenomics are still focused on coding information, leaving aside epigenomic information such as microRNAs, and methylations, among others.

6. The open pangenome of *P. aeruginosa* pathogen

As we defined in **Table 2**, an open pangenome increases when a new genome is added to the pangenome. In other words, the open pangenome is contrary to the closed pangenome, so there are changes when new genomes are added, and new strains provide new genes to the species pangenome. *P. aeruginosa was defined as an opened pangenome* [39]. *Likely, we will found continuously new strains with new genes.* As mentioned by Rouli et al. [8]: the sympatric species, present in a community, have large genomes and, thereby, an open pangenome, a high rate of HGT genes transfer, and several ribosomal operons, citing to Georgiades and Raoult [40], and Diene et al. [41]. The open pangenome of *P. aeruginosa* has a large accessory genome highly mobile with a short core genome conserved, interchanging duplication and translocations, where HGTs enter (insertions) and go out (deletions). The evolution mechanisms as

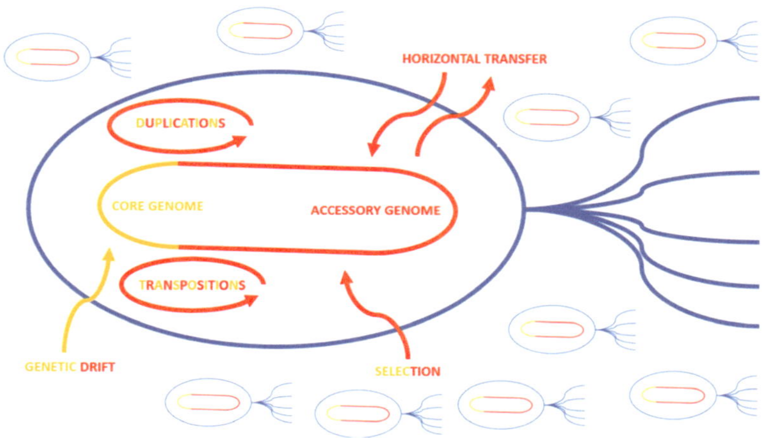

Figure 3.
Open pagenome of P. aeruginosa. The bacterial genome of P. aeruginosa has a large circular chromosome. The pangenome in total strains is divided by two regions in each chromosome, one conserved in overall strains (core genome) and the other not shared by the overall strains (accessory genome). This open pangenome among those strains has a large accessory genome highly mobile with a short core genome conserved. In the pangenome inside the species, different strains interchange HGTs, which enter (insertions) and go out (deletions). Both the core and accessory genomes undergo duplications and translocations. The evolution mechanisms such as selection, genetic drift, and others play a central role inside the genome of the strain population.

selection and genetic drift, play a central role inside the genome of the strain population (see **Figure 3**).

The *P. aeruginosa* genome has G + C content, ranging from 65 to 67%, with a size of 5.5–7 Mbp. It has a single circular chromosome and a variable number of plasmids [42]. The genome encodes a considerable repertoire of transporters, transcriptional regulators, and two-component regulatory systems, which reflects its metabolic diversity to utilize a broad range of nutrients. Additional works agree that *P. aeruginosa* has an open pangenome [39, 43]. Mosquera-Rendon et al. [4] identified distinctive positive selection in a variety of outer membrane proteins, with the data supporting the concept of genetic variation in *P. aeruginosa* proteins likely recognized as antigens. Hilker et al. [30], comparing single nucleotide polymorphism synteny indicated unrestricted gene flow between clonal complexes by recombination. Also, comparing the genomes point out that the differential genetic repertoire of clones maintains a habitat-independent gradient of virulence in the *P. aeruginosa* population.

7. The size of *P. aeruginosa* pangenome

There is a large variability in the size *of the P. aeruginosa* genome. For example, the RW109 strain has 7.049.347 bp and 6829 genes, and the PAO reference strain has 6.264.404 bp and 5697 genes. More than 700.000 bp and 1100 genes of difference between both strains, in other words, the core genome between both strains could be lower than 5697 genes, and the accessory genome could probably exceed more than 1100 genes, ignoring the repeated genes and excluding plasmids or other mobile elements. This simple exercise without robustness, and data processing without software, is less wise than Sharma et al. [3], using only five genomes, which shows the requirement to improve the level of comparison between two strains to reach the data support and confident results in pangenomics. It is the same exercise when we compare two phenotypes in metabolism or antibiotic resistance. We look by a cylinder at the wide world of the genome and its interactions, looking at a pair of genes or products, sometimes ignoring that two strains can reach the same product in two different ways. Nevertheless, Hilker et al. [30] identified 5892-7187 open reading frames (with a median of 6381 ORFs) in the *P. aeruginosa* pangenome of representative 20 strains, ranging from 6.4 to 7.4 Mbp large genomes with a core genome with approximately 4000 genes. Of course, 20 genomes is a small study for a pangenome, but there are 15 most frequent clonal complexes of the *P. aeruginosa* population.

Now, thinking about again the reported pangenomes: Sharma et al. [3] 5 genomes, Fischer et al. [5] 100 genomes, Ding et al. [7] 153 genomes, Mosquera-Rendón et al. [4] 181 genomes, and Freschi et al. [6] used 1311 genomes. It is clear that Freschi et al.'s study has the maximum number of samples and likely the best pangenome results for the *P. aeruginosa* species. However, pangenomics are not one study with the highest catalogs or number of samples. Those essential five studies have interesting matches: the accessory genome in *P. aeruginosa* is bigger than core genomes, and, thereby, *the P. aeruginosa* genome is an open pangenome, as we continue to emphasize.

Taking together the data extracted from Ding et al. [7], shown in **Figure 2** for *P. aeruginosa,* performing the same analysis for other pangenomes species, such as *Clostridium botulinum, Escherichia coli, Yersinia pestis, and Mycobacterium tuberculosis*. And taken other data from the *Mycoplasma genitalum* from Corredor & Muñoz-Gómez [10], we can compare the pangenomes of these species with each other. It is observed that *P. aeruginosa* has a smaller core genome than *Y. pestis, M. tuberculosis*, and *M. genitalum*, although greater than *E. coli* and *C. botulinum* see **Figure 4**.

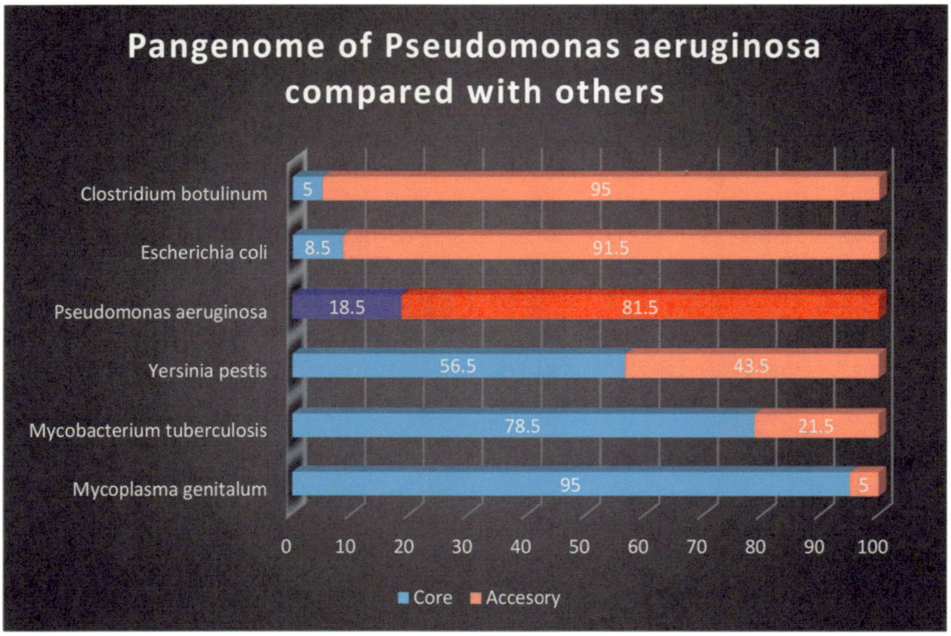

Figure 4.
Comparing the size of P. aeruginosa with other bacterial pathogens species. The comparison of P. aeruginosa pangenome (navy blue and red) with Clostridium botulinum, Escherichia coli, Yersinia pestis, Mycobacterium tuberculosis and Mycoplasma genitalum (turquoise blue and rose). Data obtained from PanX web.

8. The pangenome of MDR *P. aeruginosa*

From 2006 multidrug-resistant (MDR) *P. aeruginosa* was listed, also pandrug-resistant (PDR) [44]. The term PDR was inappropriately used in all five studies that used it and within a variety of genotypic and phenotypic characteristics. The terms MDR and PDR of *P. aeruginosa* likely cause confusion to researchers and clinicians. Falagas et al. [44] believe that at least an extended and accepted definition for the *Acinetobacter baumannii* and *P. aeruginosa* PDR should be uniformly used worldwide. Today, it is clear that MDR is a perfect term for *P. aeruginosa* antibiotic multiresistant, and probably PDR (resistance to all antibiotics) was an inconsistent term. However, the PDR acronym coins directly with pangenome, in the sense that MDR *P. aeruginosa* could be studied from pangenomics to solve the MDR problem.

The growth of MDR *P. aeruginosa* results from the extraordinary capacity of this bacterium for developing resistance through chromosomal mutations (probably in the core genome). In this species, there are increasing prevalence of transferable resistance elements as HGT genes (on the accessory and unique genome), particularly genes encoding carbapenemases or extended-spectrum β-lactamases (ESBLs) resistance. Given that *P. aeruginosa* has a nonclonal epidemic population structure and also has so much quantity of rare and unrelated genotypes that are recombining at high frequency [45].

As was mentioned previously, Freschi et al. [6], who developed the biggest pangenome of *P. aeruginosa*, initially predicted antimicrobial resistant gene profiles for 389 *P. aeruginosa* strains using the resistance gene identifier (RGI), and they found 334 unique profiles. Analysis of 1311 strains confirmed that there is, therefore, an immense variation in *P. aeruginosa* antimicrobial resistance (AMR) gene profiles, with the most frequent profile found in only 6% of the strains.

9. Antibiotic resistome

The antibiotic resistome of some organisms is the set of antibiotic-resistant genes used to resist the high concentration of some antibiotics [46]. Perhaps, the *P. aeruginosa* resistome has longtime studied [47], and now even more so with the development of next generations sequencing, databases, and the increase of pangenomics. Postgenomics data has the objective of developing the decryption of antibiotic resistome. Mercy to the other chemical techniques, today, we distinguish if the antibiotic concentration is increasing in the environment. This will have one effect on both the diversity and the abundance of resistome genes in the *P. aeruginosa* MDR. Selection for cells that carry resistance determines the addition of their relative abundance, and thus increases the more additional genes that confer resistance [48]. Fortunately, there are still susceptible strains to antibiotics, which will allow comparing those with resistant strains.

The opened pangenome of *P. aeruginosa* shows wide diversity of genes, plasmids, and mobile elements, which establish the resistome. For example, in the three reference strains, PAO1, LESB58, and PA14 are treated with seven resistant drugs exhibit phenotypic variability in their response to 27% of the antibiotics tested. For instance, the resistome annotation of 672 *P. aeruginosa* strains identified 147 loci associated with antibiotic resistance. These loci are composed mainly of acquired genomic elements and intrinsic genes [49].

10. The HGT in *P. aeruginosa* pangenome

In pangenomes of Freschi et al. [6] they study the role of HGTs in the evolution of the *P. aeruginosa* genome. They observed the flexible genes (accessory and unique genomes) and determined the proportion of these genes that were found in a single group of isolates, as well as the balance of flexible genes shared between multiple groups. Freschi et al. [6] found peculiar results because 10,515 genes (40%) of the 26,420 flexible genes were present in one group only and, on the contrary, 83 genes found in 3 groups of isolates were present in more than 90% of the isolates of a single group and, were not found in the other groups. The maximum of flexible genes was, thus, 15,905 or 60% present in isolates belonging to multiple groups; one result provides a higher estimate of the number of genes that could potentially be related to HGT episodes. They estimated the prevalence of phages because phages are often in connection with HGT events.

11. The future of pangenomics for *P. aeruginosa*

In conclusion and perspective, we can state that pangenomics had been an excellent and useful tool to decipher the *P. aeruginosa* pangenome. The open pangenome of *P. aeruginosa* displays an important diversity of HGT, mobile elements, and transposons in plasmids and phages, as a part of the accessory and unique genomes. Core genome in *P. aeruginosa* is lowest 20% of pangenome. Pangenomics is today a strategic instrument for studying the diversity in *P. aeruginosa*, useful to clarify the scene of MDR inside the specie without exception this species.

Acknowledgements

We thank the University of Antioquia for their financial support to the project CODI 2017-15753.

Conflicts of interest

The authors declare that they have no conflicts of interest regarding the publication of this paper

Author details

Mauricio Corredor*, Juan David Patiño-Salazar, Diana Carolina Castaño and Amalia Muñoz-Gómez
GEBIOMIC Group, FCEN, University of Antioquia, Medellín, Colombia

*Address all correspondence to: mauricio.corredor@udea.edu.co

References

[1] Tettelin H, Medini D. The Pangenome: Diversity, Dynamics and Evolution of Genomes. Springer; 2020. pp. 307

[2] Barh D, Soares SC, Tiwari S, Azevedo V. Pan-genomics: Applications, Challenges, and Future Prospects. Elsevier; 2020. pp. 474

[3] Sharma A, Sangwan N, Negi V, Kohli P, Khurana JP, Rao DLN, et al. Pan-genome dynamics of *Pseudomonas* gene complements enriched across hexachlorocyclohexane dumpsite. BMC Genomics. 2015;**16**(1):1-19

[4] Mosquera-Rendón J, Rada-Bravo AM, Cárdenas-Brito S, Corredor M, Restrepo-Pineda E, Benítez-Páez A. Pangenome-wide and molecular evolution analyses of the Pseudomonas aeruginosa species. BMC Genomics. 2016;**17**(1):1-14

[5] Fischer S, Klockgether J, Morán Losada P, Chouvarine P, Cramer N, Davenport CF, et al. Intraclonal genome diversity of the major *Pseudomonas aeruginosa* clones C and PA 14. Environmental Microbiology Reports. 2016;**8**(2):227-234

[6] Freschi L, Vincent AT, Jeukens J, Emond-Rheault JG, Kukavica-Ibrulj I, Dupont MJ, et al. The *Pseudomonas aeruginosa* pan-genome provides new insights on its population structure, horizontal gene transfer, and pathogenicity. Genome Biology and Evolution. 2019;**11**(1):109-120

[7] Ding W, Baumdicker F, Neher RA. panX: Pan-genome analysis and exploration. Nucleic Acids Research. 2018;**46**(1):e5-e5

[8] Rouli L, Merhej V, Fournier PE, Raoult D. The bacterial pangenome as a new tool for analysing pathogenic bacteria. New Microbes and New Infections. 2015;**7**:72-85

[9] Paralanov V, Lu J, Duffy LB, Crabb DM, Shrivastava S, Methé BA, et al. Comparative genome analysis of 19 Ureaplasma urealyticum and Ureaplasma parvumstrains. BMC Microbiology. 2012;**12**(1):1-20

[10] Corredor M, Muñoz-Gómez A. Pan-genomics of multidrug-resistant human pathogenic bacteria and their resistome. Pan-genomics: Applications, Challenges, and Future Prospects. 2020;**2020**:203-235

[11] Xiao J, Zhang Z, Wu J, Yu J. A brief review of software tools for pangenomics. Genomics, Proteomics & Bioinformatics. 2015;**13**(1):73-76

[12] Anani H, Zgheib R, Hasni I, Raoult D, Fournier PE. Interest of bacterial pangenome analyses in clinical microbiology. Microbial Pathogenesis. 2020;**149**:104275

[13] Laing C, Buchanan C, Taboada EN, Zhang Y, Kropinski A, Villegas A, et al. Pan-genome sequence analysis using Panseq: An online tool for the rapid analysis of core and accessory genomic regions. BMC Bioinformatics. 2010;**11**(1):1-14

[14] Bayjanov JR, Siezen RJ, van Hijum SA. PanCGHweb: A web tool for genotype calling in pangenome CGH data. Bioinformatics. 2010;**26**(9):1256-1257

[15] Brittnacher MJ, Fong C, Hayden HS, Jacobs MA, Radey M, Rohmer L. PGAT: A multistrain analysis resource for microbial genomes. Bioinformatics. 2011;**27**(17):2429-2430

[16] Ozer EA, Allen JP, Hauser AR. Characterization of the core and accessory genomes of Pseudomonas aeruginosa using bioinformatic tools Spine and AGEnt. BMC Genomics. 2014;**15**(1):1-17

[17] Chen X, Zhang Y, Zhang Z, Zhao Y, Sun C, Yang M, et al. PGAweb: A web server for bacterial pan-genome analysis. Frontiers in Microbiology. 2018;**9**:1910

[18] Jaiswal AK, Tiwari S, Tavares GC, da Silva WM, de Castro Oliveira L, Ibraim IC, et al. Pan-omics Focused to Crick's central dogma. In: Pan-genomics: Applications, Challenges, and Future Prospects. Editors: Barh D, Soares SC, Tiwari S, Azevedo V; 2020. pp. 1-4. Elsevier; pp. 472

[19] Medini D, Donati C, Tettelin H, Masignani V, Rappuoli R. The microbial pan-genome. Current Opinion in Genetics & Development. 2005;**15**(6): 589-594

[20] Tettelin H, Riley D, Cattuto C, Medini D. Comparative genomics: The bacterial pan-genome. Current Opinion in Microbiology. 2008;**11**(5):472-477

[21] Sigaux F. Cancer genome or the development of molecular portraits of tumors. Bulletin de L'academie Nationale de Medecine. 2000;**184**(7):1441-1447

[22] Tettelin H, Masignani V, Cieslewicz MJ, Donati C, Medini D, Ward NL, et al. Genome analysis of multiple pathogenic isolates of Streptococcus agalactiae: Implications for the microbial "pan-genome". Proceedings of the National Academy of Sciences. 2005;**102**(39):13950-13955

[23] Computational pan-genomics: Status, promises and challenges. Briefings in Bioinformatics. 2018;**19**(1):118-135

[24] Bobay LM. The prokaryotic species concept and challenges. The Pangenome. 2020:21-49

[25] Dobzhansky T. A critique of the species concept in biology. Philosophy of Science. 1935;**2**(3):344-355

[26] Mayr E. Systematics and the Origin of Species. New York: Columbia University Press; 1942, 334

[27] Hilker R, Munder A, Klockgether J, Losada PM, Tümmler B. Interclonal gradient of virulence in the Pseudomonas aeruginosa pangenome from disease and environment. Environmental Microbiology. 2014;**17**:29-46. DOI: 10.1111/1462-2920.12606

[28] Wang C, Ye Q, Jiang A, Zhang J, Shang Y, Li F, et al. *Pseudomonas aeruginosa* detection using conventional PCR and quantitative real-time PCR based on species-specific novel gene targets identified by pangenome analysis. Frontiers in Microbiology. 2022:13

[29] Baloyi IT, Adeosun IJ, Yusuf AA, Cosa S. In silico and in vitro screening of antipathogenic properties of Melianthus comosus (Vahl) against Pseudomonas aeruginosa. Antibiotics. 2021;**10**:1-23. DOI: 10.3390/antibiotics10060679

[30] Hilker R, Munder A, Klockgether J, Losada PM, Chouvarine P, Cramer N, et al. Interclonal gradient of virulence in the P seudomonas aeruginosa pangenome from disease and environment. Environmental Microbiology. 2015;**17**(1):29-46

[31] Garrity GM, Bell JA, Lilburn T. Pseudomonadales Orla-Jensen 1921, 270 AL. In: Bergey's Manual® of Systematic Bacteriology. Boston, MA: Springer; 2005. pp. 323-442

[32] Yahr TL, Parsek MR. Pseudomonas aeruginosa. In: The Prokaryotes: Vol. 6: Proteobacteria: Gamma Subclass. Editor: Martin Dworking. Springer; 2006. pp. 704-713

[33] Woese CR, Fox GE. Phylogenetic structure of the prokaryotic domain:

The primary kingdoms. Proceedings of the National Academy of Sciences of the United States of America. 1977;**74**(11):5088-5090

[34] Woese CR, Gutell R, Gupta RAMESH, Noller HF. Detailed analysis of the higher-order structure of 16S-like ribosomal ribonucleic acids. Microbiological Reviews. 1983;**47**(4):621

[35] Woese C, Kandler O, Wheelis M. Towards a natural system of organisms: Proposal for the domains Archaea, Bacteria, and Eucarya. Proceedings of the National Academy Science USA. 1990;**87**(12)

[36] Komatsoulis GA, Waterman MS. A new computational method for detection of chimeric 16S rRNA artifacts generated by PCR amplification from mixed bacterial populations. Applied and Environmental Microbiology. 1997;**63**(6):2338-2346

[37] Winand R, Bogaerts B, Hoffman S, Lefevre L, Delvoye M, Van Braekel J, et al. Targeting the 16s rRNA gene for bacterial identification in complex mixed samples: Comparative evaluation of second (illumina) and third (oxford nanopore technologies) generation sequencing technologies. International Journal of Molecular Sciences. 2019;**21**(1):298

[38] Steven B, Hesse C, Soghigian J, Gallegos-Graves LV, Dunbar J. Simulated rRNA/DNA ratios show potential to misclassify active populations as dormant. Applied and Environmental Microbiology. 2017;**83**(11):e00696-e00617

[39] Ozer EA, Nnah E, Didelot X, Whitaker RJ, Hauser AR. The population structure of Pseudomonas aeruginosa is characterized by genetic isolation of exoU+ and exoS+ lineages. Genome Biology and Evolution. 2019;**11**(7):1780-1796

[40] Georgiades K, Raoult D. Defining pathogenic bacterial species in the genomic era. Frontiers in Microbiology. 2011;**1**:151

[41] Diene SM, Merhej V, Henry M, El Filali A, Roux V, Robert C, et al. The rhizome of the multidrug-resistant Enterobacter aerogenes genome reveals how new "killer bugs" are created because of a sympatric lifestyle. Molecular Biology and Evolution. 2013;**30**(2):369-383

[42] Klockgether J, Cramer N, Wiehlmann L, Davenport CF, Tümmler B. Pseudomonas aeruginosa genomic structure and diversity. Frontiers in Microbiology. 2011;**2**:150

[43] Botelho J, Tüffers L, Fuss J, Buchholz F, Utpatel C, Klockgether J, Schulenburg H. Phylogroup-specific variation shapes pangenome dynamics in *Pseudomonas aeruginosa*. BioRxiv; 2022. pp. 1-33

[44] Falagas ME, Koletsi PK, Bliziotis IA. The diversity of definitions of multidrug-resistant (MDR) and pandrug-resistant (PDR) Acinetobacter baumannii and Pseudomonas aeruginosa. Journal of Medical Microbiology. 2006;**55**(12):1619-1629

[45] Oliver A, Mulet X, López-Causapé C, Juan C. The increasing threat of Pseudomonas aeruginosa high-risk clones. Drug Resistance Updates. 2015;**21**:41-59

[46] Wright G. The antibiotic resitome: The nexus of chemical and genetic diversity. Nature Reviews. Microbiology. 2007;**5**:175-186

[47] D'Acosta D et al. Samplig the antibiotic resitome. Science. 2006;**311**:374-377

[48] Gillings MR. Evolutionary consequences of antibiotic use for the resistome, mobilome and microbial pangenome. Frontiers in Microbiology. 2013;**4**:4

[49] Jaillard M, van Belkum A, Cady KC, Creely D, Shortridge D, Blanc B, et al. Correlation between phenotypic antibiotic susceptibility and the resistome in Pseudomonas aeruginosa. International Journal of Antimicrobial Agents. 2017;**50**(2):210-218

Chapter 6

Pseudomonas aeruginosa Represents a Main Cause of Hospital-Acquired Infections (HAI) and Multidrug Resistance (MDR)

Ehsan F. Hussein

Abstract

Pseudomonas aeruginosa is an opportunistic gram-negative bacterium, which associates with a wide range of diseases. This bacterial type is found commonly in the environment, such as soils and water. Also, it represents the main cause of hospital-acquired infections, which was associated with high rates of morbidity and mortality among aging and/or immunocompromised patients. *Ps. aeruginosa* contains virulence factors that made them able to transmit and also gave them the ability to resist many types of antibiotics. Therefore, it represents a major challenge for health workers. One of the main characterizations of *Pseudomonas aeruginosa* is antibiotic resistance; this is associated with many types of genes that are encoded for different antibiotic resistance mechanisms. The efflux pump is a very important mechanism of this bacteria. *Ps. aeruginosa* can cause a wide range of diseases, such as pneumonia, urinary tract infections, ear infections, folliculitis, puncture wounds, cystic fibrosis, burn infections, endocarditis, endophthalmitis, meningitis, and many others.

Keywords: *Ps. aeruginosa*, transmission dynamics, virulence genes, toxins production, antibiotic resistance

1. Introduction

The genus of *Pseudomonas* includes more than 140 species, most of which are saprophytic in different habitats. More than 25 species are associated with humans. Most of these bacteria are pathogenic to humans and cause opportunistic diseases. These include *Ps. aeruginosa*, *Ps. fluorescens, Ps. putida, Ps. stutzeri, Ps. cepacia, Ps. putrefaciens,* and *Ps. maltophilia*. Only two species, which are *Ps. pseudomallei* and *Ps. mallei*, produce specific human infections: glanders and melioidosis. *Ps. aeruginosa* and *Ps. maltophilia* account for approximately (80%) of pseudomonads recovered from clinical specimens. *Pseudomonas aeruginosa* represents the more frequently involved human infections and has extradited the most attention. It is widely distributed as a free-living bacterium and is found mostly in moist environments. Although it rarely causes infections in healthy persons, it is a significant threat to hospitalized patients, especially those with serious underlying diseases, such as cancers and burns. The high mortality associated with this infection is due to a combination of weak host defenses, bacterial resistance to antimicrobials, production of bacterial enzymes, and extracellular toxins [1].

Pseudomonas aeruginosa is a gram-negative opportunistic pathogen and a model bacterium for studying bacterial virulence and social traits, while it can be isolated in low numbers from a variety of environments, including soils and water, and is easily found in any human/animal-affected environment, it is a major cause of disease and death in humans with chronic conditions and immunosuppressive diseases, and the infection in these patients is difficult to treat due to a number of antibiotic resistance mechanisms and the organism's tendency to form multicellular biofilms [2]. A non-spore-forming bacterial rod is capable of causing a variety of infections in both immunocompromised and immunocompetent patients [3]. Its tendency to cause infection in immunocompromised hosts, high diversity, antibiotic resistance, and a wide range of dynamic defenses make it a very challenging organism to treat in modern medicine [4].

2. *Pseudomonas aeruginosa*

Pseudomonas aeruginosa is a heterotrophic, motile, gram-negative rod-shaped bacterium about 1–5 μm in length and 0.5–1.0 μm in wide. This bacterium is facultative aerobes that grow with aerobic and anaerobic respiration, also with nitrate as the terminal acceptor of electrons. *Ps. aeruginosa* can as well grow anaerobically *via* arginine and has limited abilities for a fermentative process that generally support quite slow or no growth. This microorganism can use more than 100 organic molecules as an origin of carbons and/or energy and as a prototroph, which generally has the capacity to grow on a medium with minimal salts and with a single origin of carbon and energy.

Pseudomonas aeruginosa is a heterotrophic, motile, rod-shaped gram-negative bacterium about 1–5 μm in length and 0.5–1.0 μm in width. It has a facultative antenna grown by aerobic and anaerobic respiration with nitrate as the terminal acceptor of electrons. *Ps. aeruginosa* also be grown anaerobic *via* arginine and has limited fermentation abilities that generally support highly slow or no growth. An organism can use more than 100 organic molecules as an origin of carbons and/or energy and as a primary source; it generally has the ability to grow on a minimal salt growth medium with a single source of carbon and energy. This bacterium grows well at 37°C, but can survive in temperature ranging from 4 to 42°C. They are important soil bacteria capable of breaking down PAHs, but they are also often detected in water tanks contaminated by animals and humans, such as sewage and drains, in and out of hospitals. The two most commonly used tested strains are PAO1 [5] and PA14, both of which were used to create genetic resources, including ordered and publicly available transposon mutant libraries. *Pseudomonas aeruginosa* is overwhelmingly resistant to several types of antibiotics, and this makes it a problem during infections where it is difficult to therapeutic processes. It is often called an "opportunistic" pathogen because it rarely infects healthy persons. In the clinic, the primary risk includes patients with compromised immune systems, including those with cystic fibrosis (CF), AIDS, cancer, static medical devices, eye injuries, burns, and nonhealing diabetic wounds.

2.1 Etiology of *Pseudomonas aeruginosa*

Pseudomonas aeruginosa is generally present in the environment, especially in freshwater. Reservoirs in urban communities include pools swimming. It can cause a large number of the society-acquired infections, such as puncture wounds leading to pneumonia, folliculitis, otitis externa, osteomyelitis, and others. It is usually an opportunistic pathogenic bacterium and is also an important cause of nosocomial infections, such as ventilator-related pneumonia, catheter-related urinary tract infections, and others. Reservoirs in hospitals include taps, drinking water, sinks, ice makers, toothbrushes, disinfection solutions, disinfectants, respiratory therapy equipment, soap bars, and endoscope washers [2–4].

2.2 Epidemiology of *Pseudomonas aeruginosa*

Ps. aeruginosa diseases are common in immunocompromised persons, such as bronchiectasis, cystic fibrosis, neutropenia, cancer, burns, organ transplant, AIDS, uncontrolled diabetes mellitus, and ICU admissions. Persons with invasive devices such as indwelling catheters and/or tubes of endotracheal are also at risk due to the organism's singular capacity to the formation of biofilms that are difficult to identify [6].

2.3 Pathophysiology of *Pseudomonas aeruginosa*

Pseudomonas aeruginosa possesses several mechanisms of antimicrobial resistance and a variety of virulence factors that collectively show the broad spectrum of infections it causes and the increasingly challenging treatment of the resulting antibiotic resistance. Multiple mechanisms of *Pseudomonas* antimicrobial resistance have been identified, for example, efflux systems, endogenous antibiotic resistance, and inactivated enzymes to antibiotics [6].

Endogenous antimicrobial resistance is the capacity to enclose membrane permeability to drugs. Flow systems allow *Ps. aeruginosa* to direct harmful or toxic materials out of the cell membrane. In addition, many of these bacterial isolates possess beta-lactamases and broad-spectrum beta-lactamases (ESBLs). The ability of these bacteria to form a biofilm is an important mechanism by which they can increase the resistance of antimicrobial agents as well against the defense mechanisms of the host [4, 6]. This is especially important in patients with cystic fibrosis, as most patients develop an infection in the first year of life either from the environment or from health care facilities (inpatient or outpatient).

2.4 History and physical of *Pseudomonas aeruginosa*

Pseudomonas aeruginosa causes a large number of infections with a variant type of severity, for example,

- Pseudofolliculitis is characterized by maculopapular rash, axillary lymphadenopathy, breast pain, and fever. It is associated with the use of hot tubs swimming pools and jacuzzis [7].

- Wounds puncture to the feet, especially in children, can lead to *pseudomonas* infection. The classic scenario is an injury to nails or glass through shoes or rubber soles. Local tenderness, purulent discharge, and other signs of infection are found; however, delay in presentation and diagnosis due to lack of symptoms initially may predispose to serious complications, such as septic arthritis and osteomyelitis [8].

- In cases of exacerbation of cystic fibrosis, sham coverage is always needed. Chronic infection with *Pseudomonas aeruginosa* occurs in more than 60% of adults with cystic fibrosis and is associated with a higher mortality rate [9].

- In persons with burns, infection of this bacterium was found to be reduced and replaced by other gram-negative and positive microorganisms, such as *Aeromonas hydrophila*. In pseudo-burn diseases, discharge of blue-green purulent may be present, among the other signs of local and systemic illness, and may lead to sepsis and septic shocks [10].

- Diabetes mellitus patients are more likely to develop malignant otitis externa caused by *Pseudomonas* spp. and present with otorrhea and otalgia. The examination often reveals granulation tissue within the ear canal [11].

- An at-risk population is at high risk for *Pseudomonas pneumonia*. Pseudomembranous pneumonia should be considered in any patient with signs and symptoms of pneumonia and in the immunocompromised patients [12].

- While patients who undergo an organ transplant are generally at risk for all types of systemic *pseudomonas* infections, kidney transplant patients are at an increased risk of developing recurrent pseudomonas infections of the urinary tract [13].

- Other examples include endocarditis, endophthalmitis, meningitis, and many other infections [14].

3. Virulence factors that cause pathogenicity of *Pseudomonas aeruginosa*

Pseudomonas aeruginosa possesses several virulence factors that may cause, pathogenesis that facilitates adhesion and/or disrupts host cell signaling pathways while targeting the extracellular matrix. This bacterium is capable of causing many infections that invade the organism and its immune system, resulting in diseases that are almost impossible to eradicate [15–17]. The virulence factors related to pathogenicity include lipopolysaccharide, type IV Pili, flagellum, type III secretion system, exotoxin A, proteases, biofilm formation, type VI secretion systems, and oxidant generation in the airspace. They are important factors that act in various ways to decrease the activity of the immune system.

The lipopolysaccharide (LPS) is a common component of the external membrane of *Ps. aeruginosa*. Bacterial lipopolysaccharide consists of a distal polysaccharide (or O-antigen), a non-repeating core oligosaccharide, and a hydrophobic domain known as lipid A (or endotoxin) [18]. LPS plays a specific role in the activations of the host's innate (non-specific) (NLRP1, NLRP2, NLRP3, and TLR4) and acquired (or adaptive) immune responses, and ends up causing dysregulated inflammatory responses that contribute to morbidity and mortality [19].

With regard to virulence factors, type IV capillaries *of Ps. aeruginosa* have a role in adhering to many cell types, and this is likely to be important in phenomena, such as tissue swelling and attachment to specific tissues. The triggering of biofilm formation and phagocytosis is mediated by phagocytic receptors that recognize the corresponding adhesions to microbial surfaces [20–24]. However, type III secretion system virulence factors provide a variety of secretion systems of which at least four likely play a role in virulence (types I, II, III, and IV). T3SS is one of the most intriguing virulence factors that involve flagella and basal body-associated system to deliver proteins directly from the cytoplasm of *Ps. aeruginosa* into the cytosol of host cells contributing to the avoidance of phagocytosis by *Ps. aeruginosa* as well as host tissue damage, promoting immune avoidance and bacterial proliferation [25, 26].

Exotoxin virulence factors are secreted by the mechanism of type II secretion, which uses proteolytic secretions in the extracellular environment, including phospholipase, lipase, protease, and alkaline phosphatase. Experiments with animals showed the important role of these factors in infectious patterns [27]. Furthermore, it has been shown that exotoxin A is involved in the damage and invasion of local tissues [28].

Several proteases that are produced by *Ps. aeruginosa* in the form of elastase Las B-type metal proteases destroy host tissues; thus, it plays an important role in both acute infections of the lung and burns wound [29]. Indeed, alginate virulence factors can produce a mucopolysaccharide capsule, consisting of alginate, an acetylated random copolymer of β1-4 bound to D-mannuronic acid (poly-M), and L-guluronic acid [30]. Evidence for playing this role is thought to be due to the overproduction of agents in CF intrapulmonary cell adhesion, and it is thought to be involved in host defense resistance by decreasing susceptibility to phagocytosis [31].

Quorum sensing virulence factor is a mechanism of the bacterial cell into cell communications with chemical compound diffusible. A quorum is necessary to produce enough of a secreted signaling molecule (called auto-stimulator) to stimulate

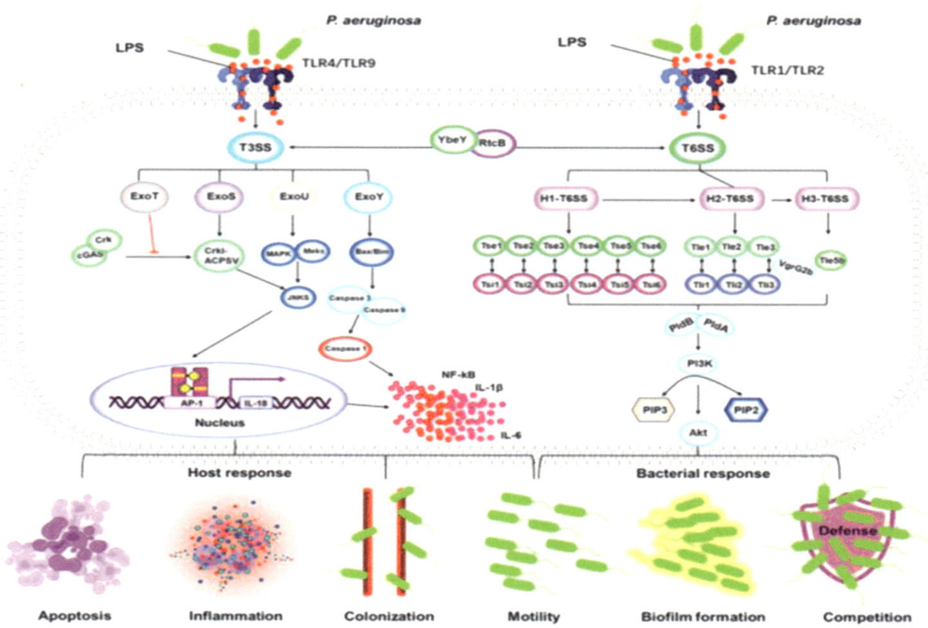

Figure 1.
Mechanisms of T3SS and T6SS in regulating bacterial pathogenesis and host responses in Ps. aeruginosa.

the expression of a large base [32]. In regard to the auto-stimulator, the class most commonly used by gram-negative bacteria and acyl homoserine lactones (AHL) is the class most commonly used. The mechanism spreads freely across bacterial membranes. The AHL signals produced by *Ps. aeruginosa* are oxohexanoyl-homoserine lactone and butanoyl-homoserine lactone [33, 34].

AHL signals produced by AHL synthase (LasI/RhlI) circulate in the environment. An increase in bacterial density during infection leads to an increase in the concentration of the auto-stimulator. At a specific moment, the autoinducer reaches a certain threshold and subsequently binds to a transcription activator (LasR/RhlR) that forms a complex that activates genes involved in biofilm formation and coding virulence factors [35]. Formation of a biofilm begins with the connection of free-swimming bacteria (plankton) to a surface with type IV filaments and skin, followed by twitching movement and formation of micro-colonies; then, the quorum sensing signals start to accumulate. Once a critical threshold for quorum detection signals is achieved, small colonies become embedded in the extracellular matrix (**Figure 1**) [15].

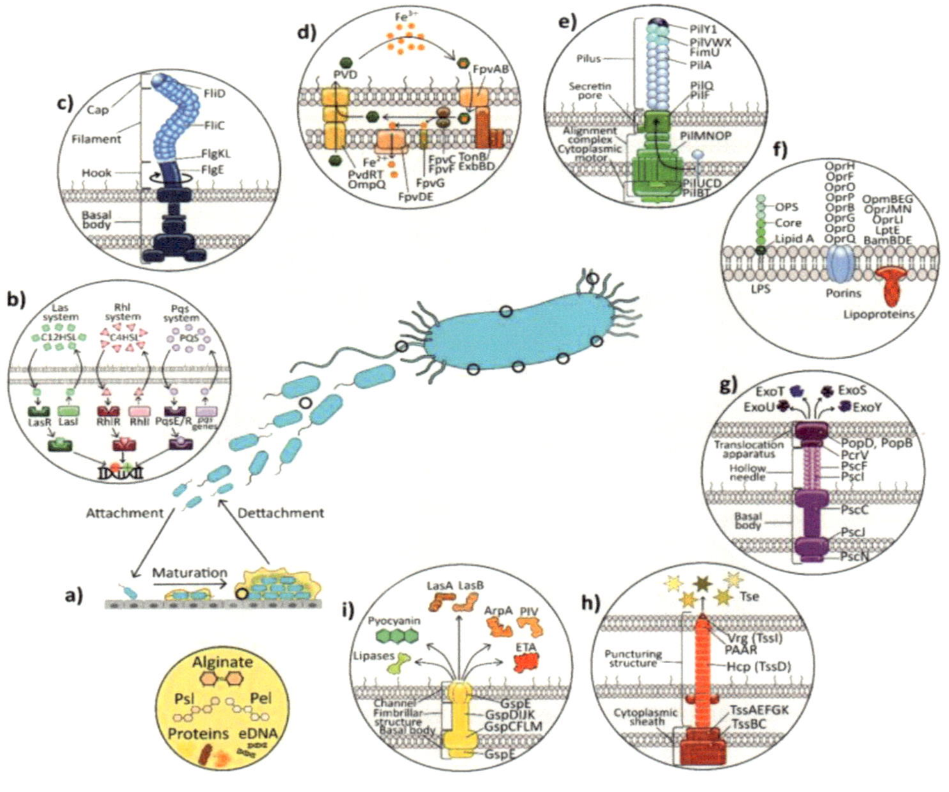

Figure 2.

Diagrammatic representation of the main virulence factors employed by Pseudomonas aeruginosa: (a) the formation capacity of the biofilm and the composition of the extracellular matrix of the biofilms (exopolysaccharides, extracellular DNA, and proteins); (b) (QSs) the three major quorum detection systems (Las, Pqs, and Rhl); (c) flagellins FliC and FliD embedded in the flagel structure; (d) pyoverdine (PVD) siderophore as important for iron uptake system; (e) (T4P) type 4 pili; (f) (LPS) lipopolysaccharide and (OMPs) outer membrane proteins; (g) (T3SS) the type III secretion system and its four main effectors; (h) (T6SS) the type VI system of secretion; (i) (T2SS) the type II system of secretion and the compounds it releases to the extracellular milieu: lytic enzymes; proteases, lipases (PIV and AprA) and elastases (LasB and LasA), (ETA) exotoxin A, and pyocyanin [36].

In this review, biofilm formation, which is a virulence factor in the first subject, is described. However, the last element of virulence discussed here is the type VI secretion system. Pathogenic bacteria often possess a number of secretion systems that function to transmit protein secretion. The T6SS is one of the more recently known examples of such secretion systems. Interest in T6SS has resulted in a rapid study of *Pseudomonas aeruginosa* in terms of structure, mechanical function, aggregation, and regulation of the secretion of *Ps. aeruginosa*. T6SS protects from other bacteria in the environment (**Figure 2**) [37, 38].

4. Genomic and phenotypic approaches to the study of *Pseudomonas aeruginosa*

Genomic and phenotypic approaches to the study of *Pseudomonas aeruginosa* adaptation in the CF lung, the ubiquity of this bacterium can be strongly related to the high plasticity of the genome. *Ps. aeruginosa* has a genome of nearly from 5.2 to 7 Mbp [39, 40], with 4000 genes inside the core genome. The complete group of genes between the different *Ps aeruginosa* strains ranges from 10,000 to 40,000 genes, and it is interesting that their arrangement in the genome may differ between strains. This makes it difficult to identify areas adapted to genetic markers [39–41]. There is large detailed information about the *Ps. aeruginosa* genome, transcriptome, and proteome available from various databases: (i) the genome database of *pseudomonas* [42]; (ii) PseudoCyc [43]; (iii) (SYSTOMONAS) [44]; (KEGG) [45]; (PubChem) [46]; and (HMDB) [47].

Advances in the sequencing of genetic systems have a possible understanding of the evolution and adaptation of *Pseudomonas aeruginosa* chronic infections, such as lung cystic fibrosis (CF), revealing high levels of coexisting genetic and phenotypic diversity, including clinically important features (**Table 1**). Whole genomic sequencing of this bacterium isolates longitudinally obtained from patients with cystic fibrosis provided evidence that, during the long term of the disease, *Ps aeruginosa* undergoes adaptive processes that lead to an assemblage of mutations in infectious strains [55].

5. Hospital-acquired infections and antibiotic resistance of *Pseudomonas aeruginosa*

Pseudomonas aeruginosa is a gram-negative bacillus belonging to the family Pseudomonadaceae [56]. In hospital environments, it can colonize damp places, including ventilators, humidifiers, oxygen respirators, sinks, taps, dialysis machines, and toilets [57]. Risk factors associated with plasmodium aeruginosa infection are COPD, diabetes, acute kidney, cystic fibrosis, immunosuppression (after organ or bone marrow transplant), liver failure, and multiple organ injury [57].

Pseudomonas infection is the most common condition treated in intensive care units (ICU), including hematology, burn units, and surgery. Clinical forms of *Ps. aeruginosa* infection are hospital-acquired pneumonia (HAP) including) urinary tract infection (UTI), ventilator-associated pneumonia (VAP and bloodstream infection (BSI), including central line-associated bloodstream infection (CLA-BSI) and wound burns, soft tissue infections, skin and surgical site infections, decubitus ulcers, eye infections, bone and joint infections, otitis media, and central nervous system infections [58].

Study type	Isolates source	Findings main	Mutated genes frequently	Identified mutated genes function	Ref.
Study of evolution *in vivo* through whole genome sequencing using.	474 clinical isolates of longitudinal cystic fibrosis from 34 adults, children and young.	36 lineages with convergent evolution in 52 genes.	*oprD, gyrA, mexA, asR, mexS, nex, gyrB, yecS, mexB, rbdA pela, and algU*	Adaptation of host, antimicrobial resistance, and loss of extracellular factors of virulence.	[48]
Evolution study *in vivo* of 17 antimicrobial resistance loci	361 independent cystic fibrosis isolates were collected from 30 cystic fibrosis centers.	1112 sequences of variants are not found in most 20 common PA clones.	*rpoB, spuE, gyrA, mexA, fusA1, mexZ, ampD, oprD, mexY, parS, envZ pagL (amgS), and parR*	Not connected. Translation, transfer, and modification of LPS and AMR.	[49]
In longitudinal analysis and *in vivo* evolution 14	Same clonal lineage isolates of cystic fibrosis (CF) patient (20 years of disease).	Toward evolution purification of choice. Evolutionary different pathways affect genes of the same functional classes.	*ampC and ftsI*	Encoding β-lactamase and penicillin-binding protein 3 (AMR)	[50]
Biofilm *in vitro* and evolution and plankton culture in stationary phase.	Study of 57 CIP-evolved populations and 35 controls.	The development of CIP resistance depends on the lifestyle of bacteria.	*ftsZ and murG sdhA*	Recycling of cell wall, arginine, TCA cycle, and catabolism.	[51]
Evolution in real time *in vivo*, a study of metabolism and genomics.	26 of one CF patient (8 years of injury).	Phenotypic level affinity but different mutation patterns.	*undefined (functional grouping)*	Transport of the amino acids, defense, metabolism, signal transductions, and translations	[52]
Analysis of genome *in vivo* (wgMLST).	Two environmental isolates, 1 veterinary, and 1 clinical CF isolate with Las QS system malfunction.	Determination of 10 highly differentiated loci between the studied strains and the PAO1 as well as PA14.	*exsA, hopJ, and rsmN*	Virulence traits regulated by T3SS and QS.	[53]

Study type	Isolates source	Findings main	Mutated genes frequently	Identified mutated genes function	Ref.
Screening of 8 illness-related phenotypes (evolution *in vivo*)	443 longitudinal isolations from 39 young patients with cystic fibrosis (CF) over 10 years.	Determine phenotypic changes that deviate from expected evolutionary paths.	*retS, mexZ, nalDmucA, nfxB, gacAS, algU, rsmA, gyrB47, and gyrA*	antibiotics flow pumps, mucosal regulators, resistance of ciprofloxacin	[54]

Abbreviations: PA, Ps. aeruginosa*; CF, cystic fibrosis; CIP, ciprofloxacin; AMR, antimicrobial resistance; LPS, lipopolysaccharide; QS, quorum sensing; TCA, tricarboxylic acid cycle; T3SS, Type 3 secretion system; wgMLST, whole genome multilocus sequence typing.*

Table 1.
Examples studies of genomic of Pseudomonas aeruginosa *adaption and evolution.*

According to the World Health Organization, *Pseudomonas aeruginosa* is one of the antimicrobial-resistant bacteria that poses the greatest threat to human health [59]. This bacterium can have several intrinsic or acquired resistance mechanisms, often with high rates of resistance to different classes of antimicrobials [60]. *Pseudomonas aeruginosa* is substantially resistant to most antibiotics due to its selective ability to prevent different antimicrobial agents move through its outer membrane or extruding if it enters the cell. The antibiotics, which remain active, include some fluoroquinolones (e.g., levofloxacin and ciprofloxacin), aminoglycosides (e.g., tobramycin, gentamicin, and amikacin), many β-lactams (e.g., ceftazidime, piperacillin-tazobactam, cefepime, ceftazidime-avibactam, imipenem, doripenem, ceftolozane-tazobactam, and meropenem), and polymyxins. Antimicrobial resistance of *Ps. aeruginosa* can be acquired through one or more mechanisms, including modified antibiotic targets, active efflux pumps, lowering of permeability, and degradation of enzymes [61].

Endogenous and acquired antibiotic resistance makes *Pseudomonas aeruginosa* and is one of the most difficult organisms to treat. The high intrinsic antibiotic resistance of *Ps. aeruginosa* is due to several mechanisms: reduced outer membrane permeability, AmpC β-lactamase production, and the presence of several genes coding for multidrug-resistant efflux pumps [62].

A large number of acquired resistance genes encoding aminoglycoside-modifying enzymes (AME) and beta-lactamases have been observed in *Ps. aeruginosa* [63]. Extended-spectrum beta-lactamases have been increasingly reported [64–67], and metallo β lactamase has also begun to appear in *Ps. aeruginosa* [68].

Mutations in chromosomal resistance genes can occur through horizontal gene transfer, and variations in the quinolone resistance determining regions (QRDR) in genes that code for topoisomerase IV and DNA gyrase play a specific role in quinolone resistance in *Pseudomonas aeruginosa* [69]. The absence of OprD is the most common mechanism of resistance to carbapenems and is linked *via* resistance to imipenem and reduced exposure to meropenems [70]. According to the data, four efflux pump systems of the RND (resistance nodulation division) family, MexEF-OprN, MexAB-OprM, MexCD-OprJ, and MexXY-OprM, are well described and known to contribute significantly to antibiotic resistance in this bacterium [71–73]. Their overproduction confers cross-resistance or reduced susceptibility to several β-lactams (piperacillin, ceftazidime, ticarcillin, meropenem, and cefepime), and aminoglycosides and quinolones [62]. Hyperproduction of the inducible AmpC β-lactamase is mostly due to the inactivation of the amidase AmpD [74] and two additional AmpD homologs [75], leading to an increase in inducer molecules.

5.1 Resistance to β-lactams of *Pseudomonas aeruginosa*

β-lactams, including penicillins (e.g., ticarcillin and piperacillin), cephalosporins (e.g., ceftazidime and cefepime), carbapenems (e.g., imipenem and meropenem), and monobactams (e.g., aztreonam), are commonly used in the treatment of *Pseudomonas aeruginosa* infections [76]. Resistance to these antibiotics is increasing [77] and mediated by different mechanisms, most commonly antimicrobial cleavage by β-lactamase enzymes, and antimicrobial expulsion by chromosomally encoded efflux pump mechanisms and reduced antibiotics uptake owing to loss of outer membrane porin proteins [78].

5.1.1 β-lactamases

β-lactamases, hydrolyzing enzymes that disrupt the amide bond of the traditional four-membered β-lactam ring, rendering antimicrobials ineffective, are the main identification of resistance in bacteria type gram-negative, such as *Ps. aeruginosa*. Four molecular classes of these enzymes (A–D) have been characterized, which include metal-dependent ($Zn2+$-requiring; class B) and metal-independent (active site serine; classes A, C, and D-lactamases reviewed, all these enzymes were reported in *Pseudomonas aeruginosa*) [79].

5.1.1.1 Endogenous β-lactamases

Ps. aeruginosa typically carries chromosomal genes for two β-lactamases, a class C cephalosporinase, AmpC (Lodge et al., 1990), and a class D oxacillinase, PoxB [80]. AmpC is a well-characterized β-lactamase commonly linked to β-lactam resistance in clinical isolates.

5.1.1.2 Acquired β-lactamases

Whereas the original β-lactamases were plasmid-encoded class A restriction enzymes that hydrolyze only older narrow-spectrum penicillins and cephalosporins, recently described β-lactamases in P. classes A and D capable of hydrolyzing a wide range of β-lactams, including broad-spectrum cephalosporins and monobactams, and carbapenemases (classes A, B, and D) that hydrolyze most β-lactams, include carbapenems, but not aztreonam [79].

5.1.1.3 Spectrum extended of β-lactamases

Reported more commonly in the *Enterobacteriaceae*, though also found in *Pseudomonas aeruginosa*, ESBLs typically hydrolyze and, so, provide resistance to broad-spectrum cephalosporins (e.g., the third generation of cefotaxime, oxyimino-cephalosporins, and ceftazidime) and azithromycin, in addition to narrow-spectrum cephalosporins and penicillins [81].

5.1.1.4 Carbapenemases

Carbapenems (e.g., meropenem and imipenem) are an important class of pseudo-beta-lactams due to their stability to most β-lactamases and are of particular use in treating infections associated with ESBL and AmpC producers [82].

5.1.2 Fluoroquinolones resistance of Pseudomonas aeruginosa

Fluoroquinolones (FQs), particularly ciprofloxacin, are generally used in the treatment of *Ps. aeruginosa* diseases. Resistance against these agents, which are mediated by mutations in the enzymes of topoisomerase IV and DNA gyrase that are the targets of FQs, although flux is an important contributing factor [83], is often accompanied by target site mutations [84].

5.1.3 Resistance to aminoglycosides

A number of aminoglycosides are commonly used in the treatment of *Ps. aeruginosa* infections (e.g., tobramycin, amikacin, and gentamicin [85], especially pulmonary infections in cystic fibrosis CF patients where amikacin and, in particular, tobramycin are routinely used [86]) However, their use is associated with the development of resistance, with acquired aminoglycoside-modifying enzymes (AMEs) and rRNA methylases, the inflow mechanisms usually responsible.

5.1.4 Biofilm resistance of Pseudomonas aeruginosa

Biofilms, which are surface-mounted three-dimensional structures in which bacteria are inserted into a matrix consisting of polysaccharides, protein, and DNA, are increasingly recognized as the preferred method for bacterial growth in nature and infectious diseases [87]. This is true for *Ps. aeruginosa* [88].

5.2 Efflux pumps of *Pseudomonas aeruginosa*

In *Ps. aeruginosa*, four efflux pumps have been described, MexCD-OprJ, MexEF-OprN, MexAB-OprM, and MexXY [89]. The genes encoding these pumps are arranged as operons, with the first gene encoding a membrane fusion protein that is associated with the cytoplasmic membrane (MexC, MexA, MexX, and MexE). The second gene encodes the transporter (MexF, MexD MexY, and MexB) thought to export the substrate across the inner membrane. The third gene encodes an OM protein (OprM, OprN, and OprJ) that facilitates the passage of the substrate across the OM. Together, the three pump proteins form a channel that traverses the inner membrane and the OM and allows the target to be effluxed directly from the cytoplasm to the extracellular environment. In the case of the*mexXY* operon, there is no gene encoding the OM component; rather, MexXY appears to share the OprM channel with MexAB [90]. However, it suggests that OprM is not the OM channel for AmrAB (MexXY) [91].

As is evident from this, there is a high variation in the antibiotic susceptibility pattern against these pathogenic bacteria, and the reason for this might be linked to an antibiotics manufacturing company, purchasing without a proper prescription, prescription without laboratory guidance, misuse, and indiscriminate use, the study area and bacterial type, and resistance methods [92–95].

Pigments of blue and green, such as (Pyocyanin), as antimicrobial agents were produced by marine *Ps. aeruginosa*. Also, the technique of molecular biology was applied to know the pathways of this production. The produced pyocyanin pigments were used as antimicrobial agents against a wide range of pathogenic microorganisms [96–99].

6. Conclusions

Pseudomonas aeruginosa is an opportunistic gram-negative bacterium and causes hospital-acquired infections, which associate with a wide range of diseases, such as pneumonia, urinary tract infections, ear infections, folliculitis, puncture wounds, cystic fibrosis, burn infections, endocarditis, endophthalmitis, meningitis, and many others. *Ps. aeruginosa* contains virulence factors that made them able to transmit and also gave them the ability to resist many types of antibiotics; this is associated with many types of genes which are encoded for different antibiotic resistance mechanisms. The efflux pumps is very important mechanism of this bacteria. *Ps. aeruginosa* can cause a wide range of diseases.

Conflict of interest

The authors declare no conflict of interest

Notes/thanks/other declarations

Thanks

Author details

Ehsan F. Hussein
University of Babylon, College of Medical Hammurabi, Babylon, Iraq

*Address all correspondence to: ehsan.f.hussein@gmail.com

References

[1] Iglewski BH. Pseudomonas. In: Baron S, editor. Medical Microbiology. 4th ed. Chapter 27. Galveston: University of Texas Medical Branch at Galveston; 1996

[2] Diggle SP, Whiteley M. Microbe profile: *Pseudomonas aeruginosa*: Opportunistic pathogen and lab rat. Microbiology. 2020;**166**(1):30-33

[3] Kerr KG, Snelling AM. *Pseudomonas aeruginosa*: A formidable and ever-present adversary. The Journal of Hospital Infection. 2009;**73**(4):338-344

[4] Mulcahy LR, Isabella VM, Lewis K. *Pseudomonas aeruginosa* biofilms in disease. Microbial Ecology. 2014;**68**(1):1-12

[5] Stover CK, Pham XQ, Erwin AL, Mizoguchi SD, Warrener P, et al. Complete genome sequence of *Pseudomonas aeruginosa* PAO1, an opportunistic pathogen. Nature. 2000;**406**:959-964

[6] Pang Z, Raudonis R, Glick BR, Lin TJ, Cheng Z. Antibiotic resistance in *Pseudomonas aeruginosa*: Mechanisms and alternative therapeutic strategies. Biotechnology Advances. 2019;**37**(1): 177-192

[7] Zacherle BJ, Silver DS. Hot tub folliculitis: A clinical syndrome. West Journal of Medicine. 1982;**137**(3):191-194

[8] Niall DM, Murphy PG, Fogarty EE, Dowling FE, Moore DP. Puncture wound related pseudomonas infections of the foot in children. Irish Journal of Medical Science. 1997;**166**(2):98-101

[9] Lund-Palau H, Turnbull AR, Bush A, Bardin E, Cameron L, Soren O, et al. *Pseudomonas aeruginosa* infection in cystic fibrosis: Pathophysiological mechanisms and therapeutic approaches. Expert Review of Respiratory Medicine. 2016;**10**(6):685-697

[10] Azzopardi EA, Azzopardi SM, Boyce DE, Dickson WA. Emerging gram-negative infections in burn wounds. Journal of Burns Care Research. 2011;**32**(5):570-576

[11] Slattery WH, Brackmann DE. Skull base osteomyelitis. Malignant external otitis. Otolaryngologic Clinics of North America. 1996;**29**(5):795-806

[12] Makhnevich A, Feldhamer KH, Kast CL, Sinvani L. Aspiration pneumonia in older adults. Journal of Hospital Medicine. 2019;**14**(7):429-435

[13] Rivera-Sanchez R, Delgado-Ochoa D, Flores-Paz RR, García-Jiménez EE, Espinosa-Hernández R, Bazan-Borges AA, et al. Prospective study of urinary tract infection surveillance after kidney transplantation. BMC Infectious Diseases. 2010;**10**:245

[14] Bodey GP, Bolivar R, Fainstein V, Jadeja L. Infections caused by *Pseudomonas aeruginosa*. Reviews of Infectious Diseases. 1983;**5**(2):279-313

[15] Skariyachan S, Sridhar VS, Packirisamy S, Kumargowda ST, Challapilli SB. Recent perspectives on the molecular basis of biofilm formation by *Pseudomonas aeruginosa* and approaches for treatment and biofilm dispersal. Folia Microbiologica. 2018;**63**:413-432

[16] Laverty G, Gorman SP, Gilmore BF. Biomolecular mechanisms of *Pseudomonas aeruginosa* and *Escherichia coli* biofilm formation. Pathogens. 2014;**3**:596-632

[17] Lambert PA. Mechanisms of antibiotic resistance in *Pseudomonas*

aeruginosa. Journal of Royal Society Medicine. 2002;**41**:22-26

[18] Pier GB, Ramphal R. *Pseudomonas aeruginosa*. In: Mandell GL, Bennett JE, Dolin R, editors. Mandell, Douglas and Bennett's Principles and Practice of Infectious Disease. 6th ed. New York: Churchill Livingstone; 2005. pp. 2587-2615

[19] Mandell G, Bennett J, Dolin R. Principles and Practice of Infectious Diseases. 7th ed. New York: Elsevier/ Churchill Livingstone; 2009

[20] Raetz CR, Whitfield C. Lipopolysaccharide endotoxins. Annual Review of Biochemistry. 2007;**71**:635-700

[21] Köhler T, Curty LK, Barja F, Van Delden C, Pechère JC. Swarming of *Pseudomonas aeruginosa* is dependent on cell-to-cell signaling and requires flagella and pili. Journal of Bacteriology. 2000;**21**:5990-5996

[22] Doring G, Pier GB. Vaccines and immunotherapy against *Pseudomonas aeruginosa*. Vaccine. 2008;**8**:1011-1024

[23] Barken KB, Pamp SJ, Yang L, Gjermansen M, Bertrand JJ, Klausen M, et al. Roles of type IV pili, flagellum-mediated motility and extracellular DNA in the formation of mature multicellular structures in Pseudomonas aeruginosa biofilms. Environmental Microbiology. 2008;**9**:2331-2343

[24] Mahenthiralingam E, Speert DP. Nonopsonic phagocytosis of *Pseudomonas aeruginosa* by macrophages and polymorphonuclear leukocytes requires the presence of the bacterial flagellum. Infection and Immunity. 1995;**11**:4519-4523

[25] Punsalang AP, Sawyer WD. Role of pili in the virulence of Neisseria gonorrhoeae. Infection and Immunity. 1973;**2**:255-263

[26] Frank DW. The exoenzyme S regulon of *Pseudomonas aeruginosa*. Molecular Microbiology. 1997;**26**:621-629

[27] Passador L, Iglewski W. ADP-ribosylating toxins. Methods in Enzymology. 1994;**235**:617-631

[28] Iglewski BH, Liu PV, Kabat D. Mechanism of action of *Pseudomonas aeruginosa* exotoxin Aiadenosine diphosphate-ribosylation of mammalian elongation factor 2 in vitro and in vivo. Infection and Immunity. 1977;**1**:138-144

[29] Bielecki P, Glik J, Kawecki M, Martins dos Santos VA. Towards understanding Pseudomonas aeruginosa burn wound infections by profiling gene expression. Biotechnology Letters. May 2008;**30**(5):777-790

[30] Pier GB, Coleman F, Grout M, Franklin M, Ohman DE. Role of alginate O acetylation in resistance of mucoid Pseudomonas aeruginosa to opsonic phagocytosis. Infection and Immunity. 2001;**3**:1895-1901

[31] Diggle SP, Cornelis P, Williams P, Camara M. 4-quinolone signalling in *Pseudomonas aeruginosa*: Old molecules, new perspectives. International Journal of Medical Microbiology. 2006;**2-3**:83-91

[32] Pearson JP, Gray KM, Passador L, Tucker KD, Eberhard A, Iglewski BH, et al. Structure of the autoinducer required for expression of *Pseudomonas aeruginosa* virulence genes. Proceedings of the National Academy of Sciences. 2001;**1**:197-201

[33] Pearson JP, Passador L, Iglewski BH, Greenberg EP. A second N-acylhomoserine lactone signal produced by *Pseudomonas aeruginosa*. Proceedings of the National Academy of Sciences. 1995;**5**:1490-1494

[34] Hirakawa H, Tomita H. Interference of bacterial cell-to-cell communication:

A new concept of antimicrobial chemotherapy breaks antibiotic resistance. Frontiers in Microbiology. 2013;**114**:1-14

[35] Hauser AR. The type III secretion system of *Pseudomonas aeruginosa*: Infection by injection. Nature Reviews Microbiology. 2009;**9**:654-665

[36] Jurado-Martín I, Sainz-Mejías M, McClean S. *Pseudomonas aeruginosa*: An audacious pathogen with an adaptable arsenal of virulence factors. International Journal of Molecular Sciences. 2021;**22**:3128

[37] Ho BT, Dong TG, Mekalanos JJ. A view to a kill: The bacterial type VI secretion system. Cell Host & Microbe. 2014;**1**:9-21

[38] Miguel JC, Costa JC, Espeschit IF, PieriI FA, Benjamin MA, Moreira C. Increase in biofi lm formation by Escherichia coli under conditions that mimic the mastitic mammary gland. Ciência Rural. 2014;**4**:666-671

[39] Varga JJ, Barbier M, Mulet X, Bielecki P, Bartell JA, Owings JP, et al. Genotypic and phenotypic analyses of a Pseudomonas aeruginosa chronic bronchiectasis isolate reveal differences from cystic fibrosis and laboratory strains. BMC Genomics. 2015;**16**:883

[40] Mielko KA, Jabłonski SJ, Milczewska J, Sands D, Łukaszewicz M, Młynarz P. Metabolomic studies of *Pseudomonas aeruginosa*. World Journal of Microbiology and Biotechnology. 2019;**35**:178

[41] Parkins MD, Somayaji R, Waters VJ. Epidemiology, biology, and impact of clonal *Pseudomonas aeruginosa* infections in cystic fibrosis. Clinical Microbiology Reviews. 2018;**31**

[42] Winsor GL, Griffiths EJ, Lo R, Dhillon BK, Shay JA, Brinkman FS. Enhanced annotations and features for comparing thousands of Pseudomonas genomes in the Pseudomonas genome database. Nucleic Acids Research. 2016;**44**:D646-D653

[43] Romero P, Karp P. PseudoCyc, a pathway-genome database for Pseudomonas aeruginosa. Journal of Molecular Microbiology and Biotechnology. 2003;**5**:230-239

[44] Choi C, Münch R, Leupold S, Klein J, Siegel I, Thielen B, et al. SYSTOMONAS– an integrated database for systems biology analysis of Pseudomonas. Nucleic Acids Research. 2007;**35**:D533-D537

[45] Kanehisa M, Furumichi M, Tanabe M, Sato Y, Morishima K. KEGG: New perspectives on genomes, pathways, diseases and drugs. Nucleic Acids Research. 2017;**45**:D353-D361

[46] Kim S, Thiessen PA, Bolton EE, Chen J, Fu G, Gindulyte A, et al. PubChem substance and compound databases. Nucleic Acids Research. 2016;**44**:D1202-D1213

[47] Wishart DS, Jewison T, Guo AC, Wilson M, Knox C, Liu Y, et al. HMDB 3.0—The human metabolome database in 2013. Nucleic Acids Research. 2013;**41**:D801-D807

[48] Marvig RL, Sommer LM, Molin S, Johansen HK. Convergent evolution and adaptation of *Pseudomonas aeruginosa* within patients with cystic fibrosis. Nature Genetics. 2015;**47**:57-64

[49] Greipel L, Fischer S, Klockgether J, Dorda M, Mielke S, Wiehlmann L, et al. Molecular epidemiology of mutations in antimicrobial resistance loci of *Pseudomonas aeruginosa* isolates from airways of cystic fibrosis patients. Antimicrobial Agents Chemotherapy. 2016;**60**:6726-6734

[50] Colque CA, Albarracín Orio AG, Feliziani S, Marvig RL, Tobares AR,

Johansen HK, et al. Hypermutator *Pseudomonas aeruginosa* exploits multiple genetic pathways to develop multidrug resistance during long-term infections in the airways of cystic fibrosis patients. Antimicrobial Agents Chemotherapy. 2020;**64**:e02142-e02119

[51] Ahmed MN, Abdelsamad A, Wassermann T, Porse A, Becker J, Sommer MOA, et al. The evolutionary trajectories of *Ps. aeruginosa* in biofilm and planktonic growth modes exposed to ciprofloxacin: Beyond selection of antibiotic resistance. NPJ Biofilms Microbiomes. 2020;**6**:28

[52] La Rosa R, Johansen HK, Molin S. Convergent metabolic specialization through distinct evolutionary paths in *Pseudomonas aeruginosa*. mBio. 2018;**2018**:9

[53] Martínez-Carranza E, García-Reyes S, González-Valdez A, Soberón-Chávez G. Tracking the genome of four Pseudomonas aeruginosa isolates that have a defective Las quorum-sensing system, but are still virulent. Access Microbiology. 2020;**2**:acmi000132

[54] Bartell JA, Sommer LM, Haagensen JAJ, Loch A, Espinosa R, Molin S, et al. Evolutionary highways to persistent bacterial infection. Nature Communications. 2019;**10**:629

[55] Rossi E, La Rosa R, Bartell JA, Marvig RL, Haagensen JAJ, Sommer LM, et al. *Pseudomonas aeruginosa* adaptation and evolution in patients with cystic fibrosis. Nature Reviews Microbiology. May 2021;**19**(5):331-342

[56] Alhazmi A. *Pseudomonas aeruginosa* – Pathogenesis and pathogenic mechanism. International Journal of Biology. 2015;**7**:44-67

[57] Bassetti M, Vena A, Croxatto A, Righi E, Guery B. How to manage *Pseudomonas aeruginosa* infections. Drugs Context. 2018;**7**:212527

[58] Moore NM, Flaws ML. Epidemiology and pathogenesis of *Pseudomonas aeruginosa* infection. Clinical Laboratory Science. 2011;**24**(1):43-46

[59] Dey R, Rieger AM, Stephens C, Ashbolt NJ. Interactions of *Pseudomonas aeruginosa* with *Acanthamoeba polyphaga* observed by imaging flow cytometry. Cytometry. 2019;**95**:555-564

[60] Reygaert WC. An overview of the antimicrobial resistance mechanisms of bacteria. AIMS Microbiology. 26 Jun 2018;**4**(3):482-501

[61] Pachori P, Gothalwal R, Gandhi P. Emergence of antibiotic resistance *Pseudomonas aeruginosa* in intensive care unit; a critical review. Genes & Diseases. 17 Apr 2019;**6**(2):109-119

[62] Livermore DM. Multiple mechanisms of antimicrobial resistance in *Pseudomonas aeruginosa*: Our worst nightmare? Clinical Infectious Diseases. 2002;**34**:634-640

[63] Bonomo RA, Szabo D. Mechanisms of multidrug resistance in *Acinetobacter* species and *Pseudomonas aeruginosa*. Clinical Infectious Diseases. 2006;**43**:S49-S56

[64] Castanheira M, Mendes RE, Walsh TR, Gales AC, Jones RN. Emergence of the extended-spectrum β-lactamase GES-1 in a *Pseudomonas aeruginosa* strain from Brazil: Report from the SENTRY antimicrobial surveillance program. Antimicrobial Agents Chemotherapy. 2004;**48**:2344-2345

[65] Girlich D, Naas T, Leelaporn A, Poirel L, Fennewald M, Nordmann P. Nosocomial spread of the integron-located veb-1-like cassette encoding an extended-spectrum β-lactamase in *Pseudomonas aeruginosa* in Thailand. Clinical Infectious Diseases. 2002;**34**:603-611

[66] Poirel L, Lebessi E, Castro M, Fevre C, Foustoukou M, Nordmann P. Nosocomial outbreak of extended-spectrum β-lactamase SHV-5-producing isolates of *Pseudomonas aeruginosa* in Athens, Greece. Antimicrobial Agents Chemotherapy. 2004;**48**:2277-2279

[67] Vahaboglu H, Ozturk R, Aygun G, Coskunkan F, Yaman A, Kaygusuz A, et al. Widespread detection of PER-1-type extended-spectrum β-lactamases among nosocomial *Acinetobacter* and *Pseudomonas aeruginosa* isolates in Turkey: A nationwide multicenter study. Antimicrobial Agents Chemotherapy. 1997;**41**:2265-2269

[68] Walsh TR, Toleman MA, Poirel L, Nordmann P. Metallo-β-lactamases: The quiet before the storm? Clinical Microbiology Reviews. 2005;**18**:306-325

[69] Jalal S, Wretlind B. Mechanisms of quinolone resistance in clinical strains of *Pseudomonas aeruginosa*. Microbial Drug Resistance. 1998;**4**:257-261

[70] Ochs MM, McCusker MP, Bains M, Hancock RE. Negative regulation of the *Pseudomonas aeruginosa* outer membrane porin OprD selective for imipenem and basic amino acids. Antimicrobial Agents Chemotherapy. 1999;**43**:1085-1090

[71] Koehler T, Michea-Hamzehpour M, Henze U, Gotoh N, Curty LK, Pechere JC. Characterization of MexE-MexF-OprN, a positively regulated multidrug efflux system of *Pseudomonas aeruginosa*. Molecular Microbiology. 1997;**23**:345-354

[72] Mine T, Morita Y, Kataoka A, Mizushima T, Tsuchiya T. Expression in *Escherichia coli* of a new multidrug efflux pump, MexXY, from *Pseudomonas aeruginosa*. Antimicrobial Agents Chemotherapy. 1999;**43**:415-417

[73] Poole K, Gotoh N, Tsujimoto H, Zhao Q, Wada A, Yamasaki T, et al. Overexpression of the *mexC-mexD-oprJ* efflux operon in *nfxB*-type multidrug-resistant strains of *Pseudomonas aeruginosa*. Molecular Microbiology. 1996;**21**:713-724

[74] Juan C, Macia MD, Gutierrez O, Vidal C, Perez JL, Oliver A. Molecular mechanisms of β-lactam resistance mediated by AmpC hyperproduction in *Pseudomonas aeruginosa* clinical strains. Antimicrobial Agents Chemotherapy. 2005;**49**:4733-4738

[75] Juan C, Moya B, Perez JL, Oliver A. Stepwise upregulation of the *Pseudomonas aeruginosa* chromosomal cephalosporinase conferring high-level β-lactam resistance involves three AmpD homologues. Antimicrobial Agents Chemotherapy. 2006;**50**:1780-1787

[76] Paul M, Yahav D, Bivas A, Fraser A, Leibovici L. Antipseudomonal β-lactams for the initial, empirical, treatment of febrile neutropenia: Comparison of β-lactams. Cochrane Database of Systematic Reviews. 2010;**11**:CD005197

[77] Zilberberg MD, Chen J, Mody SH, Ramsey AM, Shorr AF. Imipenem resistance of *Pseudomonas* in pneumonia: A systematic literature review. BMC Pulmonary Medicine. 2010;**10**:45

[78] Pfeifer Y, Cullik A, Witte W. Resistance to cephalosporins and rbapenems in Gram-negative bacterial pathogens. International Journal of Medical Microbiology. 2010;**300**:371-379

[79] Zhao WH, Hu ZQ. β-lactamases identified in clinical isolates of *Pseudomonas aeruginosa*. Critical Review in Microbiologica. 2010;**36**:245-258

[80] Kong KF, Jayawardena SR, Del Puerto A, Wiehlmann L, Laabs U, Tummler B, et al. Characterization of poxB, a chromosomal-encoded

Pseudomonas aeruginosa oxacillinase. Gene. 2005;**358**:82-92

[81] Paterson DL, Bonomo RA. Extended-spectrum β-lactamases: A clinical update. Clinical Microbiology Reviews. 2005;**18**:657-686

[82] El Gamal MI, Oh CH. Current status of carbapenem antibiotics. Current Topics in Medicinal Chemistry. 2010;**10**:1882-1897

[83] Jacoby GA. Mechanisms of resistance to quinolones. Clinical Infectious Diseases. 2005;**41**(Suppl. 2):S120-S126

[84] Higgins PG, Fluit AC, Milatovic D, Verhoef J, Schmitz FJ. Mutations in GyrA, ParC, MexR and NfxB in clinical isolates of *Pseudomonas aeruginosa*. International Journal of Antimicrobial Agents. 2003;**21**:409-413

[85] Gilbert DN, Moellering RC Jr, Sande MA. The Sanford Guide to Antimicrobial Therapy 2003. Hyde Park: Antimicrobial Therapy, Inc; 2003

[86] Canton R, Cobos N, de Gracia J, Baquero F, Honorato J, Gartner S, et al. Antimicrobial therapy for pulmonary pathogenic colonisation and infection by Pseudomonas aeruginosa in cystic fibrosis patients. Clinical Microbiology Infections. 2005;**11**:690-703

[87] Lopez D, Vlamakis H, Kolter R. Biofilms Cold Springer Harbour Perspective Biology. Biofilms Cold Springer Harbour Perspective Biology. 2010;**2**:a000398

[88] Harmsen M, Yang L, Pamp SJ, Tolker-Nielsen T. An update on *Pseudomonas aeruginosa* biofilm formation, tolerance, and dispersal. FEMS Immunology and Medical Microbiology. 2010;**59**:253-268

[89] Li XZ, Nikaido H, Poole K. Role of MexA-MexB-OprM in antibiotic efflux in *Pseudomonas aeruginosa*. Antimicrobial Agents and Chemotherapy. 1995;**39**:1948-1953

[90] Aires JR, Kohler T, Nikaido H, Plesiat P. Involvement of an active efflux system in the natural resistance of *Pseudomonas aeruginosa* to aminoglycosides. Antimicrobial Agents and Chemotherapy. 1999;**43**:2624-2628

[91] Westbrock-Wadman S, Sherman DR, Hickey MJ, Coulter SN, Zhu YQ, Warrener P, et al. Characterization of a *Pseudomonas aeruginosa* efflux pump contributing to aminoglycoside impermeability. Antimicrobial Agents and Chemotherap. 1999;**43**:2975-2983

[92] Hussein EF. Estimation of the antibiotic activity against *Pseudomonas* spp isolated from ear infection. The Journal of Communicable Diseases. 2021;**53**(3):227-231

[93] Hussein EF. Detection of the antibiotic susceptibility against proteus species and *Escherichia coli* isolated from patients with ear infections. International Journal of Drug Delivery Technology. 2022;**12**(1):221-224

[94] Hussein EF, Ameen JA, Yassen SH. Study the antibiotics activity against *Escherichia Coli* isolated from urine samples of pregnant women with urinary tract infection. International Journal of Pharmaceutical Research. 2021;**13**(1):1368-1372

[95] Mousa BA, Hussein EF. Estimation the antibiotics activity against vaginal pathogenic microorganisms in pregnant ladies with early rupture membrane and their fetal outcome. Indian Journal of Forensic Medicine and Toxicology. 2020;**14**(1):455-461

[96] Darwesh OM, Barakat KM, Mattar MZ, Sabae SZ, Hassan SH.

Production of antimicrobial blue green pigment Pyocyanin by marine *Pseudomonas aeruginosa*. Biointerface Research in Applied Chemistry. 2019;**9**(5):4334-4339

[97] Li H, Wang H, Darwesh OM, Du J, Liu S, Li C, et al. Separation of biobutanol from ABE fermentation broth using lignin as adsorbent: A totally sustainable approach with effective utilization of lignocellulose. International Journal of Biological Macromolecules. 2021;**174**:11-21

[98] Xue X, Gu Q, Habimana P, Darwesh OM, Zhang B, Li Z. Simulation and optimization of three-column triple-effect methanol distillation scheme. Chemical Engineering and Processing Process Intensification. 2021;**159**:108229

[99] Goranson J, Hovey AK, Frank DW. Functional analysis of exsC and exsB in regulation of exoenzyme S production by *Pseudomonas aeruginosa*. Journal of Bacteriology.**5**:1646-1654